Glimpses of Cotton Breeding

Problems and Achievements

The Authors

Dr. Phundan Singh [b. 1946] hails from a reputed agricultural family of Western Uttar Pradesh [Village-Sakauti, Tehsil- Mawana, District- Meerut]. He graduated in Agriculture from Agra University, Agra and obtained his Master and Doctorate degrees from Kanpur University. He has throughout a brilliant academic record. His area of specialization is Plant Breeding and Genetics. He was selected as Scientist in first batch of ARS examination in 1976.

Dr. P. Singh has over 40 year's research experience of Plant Breeding and Genetics. He has served Central Institute for Cotton Research, Nagpur in various capacities such as Scientist, Senior Scientist, Principal Scientist, Head, Division of Crop Improvement and Director. He has 300 publications to his credit. He has participated and presented 80 research papers in various National and International Seminars, Symposia and Conferences. He has authored 57 books on Genetics and Plant Breeding mostly published by Kalyani Publishers, New Delhi. He has also authored 18 technical bulletins on various aspects of cotton and contributed 14 chapters in various books.

Dr P. Singh is a member of several Scientific Societies, referee of different journals, Expert member of Selection Committees and M. Sc. and Ph. D. Examiner in different Agricultural Universities. He is also paper setter of B. Sc. and M. Sc. in different Universities. He has served as the Chief Editor of Hybrid Cotton Newsletter for over 10 years at CICR. The institute received the best annual report award and was recognized as *Bt.* referral laboratory in 2003 when he was the Director. He has visited several countries such as Russia, Belarus, Ukraine, Crimea, Uzbekistan, Canada, USA and England.

Sanjeev Singh obtained his post graduate degree in Botany from Nagpur University, Nagpur in first division. His area of specialization is Genetics and Plant Breeding. His M.Sc. dissertation was "Exploitation of Heterosis in Cotton: A Review". Immediately after postgraduation he joined as a teacher in Junior College and started teaching Botany. After three years he joined as a Sales Manager in a Pharmaceutical Company. Later on he was elevated to the post of Area Manager (Vidarbha) and Regional Manager (Maharashtra and Goa). He has authored three books including present one. His other books are Heterosis Breeding in Cotton and Breeding Hybrid Cotton. Unfortunately he left the planet at a very tender age.

His hobbies were playing cricket and chess, body building and site seeing *etc.* He participated in several regional and national level chess competitions and obtained certificates. He organized some chess tournaments at national level and was member of a monthly magazine Chess Mate.

Glimpses of Cotton Breeding

Problems and Achievements

Phundan Singh

Sanjeev Singh

2018

Daya Publishing House®

A Division of

Astral International Pvt. Ltd.

New Delhi – 110 002

© 2018 AUTHORS

ISBN 9789387057630(International Edition)

Published by : **Daya Publishing House**®
A Division of
Astral International Pvt. Ltd.
– ISO 9001:2015 Certified Company –
4736/23, Ansari Road, Darya Ganj
New Delhi-110 002
Ph. 011-43549197, 23278134
E-mail: info@astralint.com
Website: www.astralint.com

Dedicated to Children of Junior Author
Er. Miss Shubhika Singh
Mr. Rishabh Singh

Preface

In India, remarkable work has been done on cotton breeding in different cotton growing states and several high yielding varieties and hybrids have been developed for different regions addressing to location specific problems. There are several books on this subject both of foreign and Indian origin. However, there is hardly any book which provides comprehensive information on cotton breeding in India. The prime objective of writing the book entitled: Glimpses of Cotton Breeding" is to fill up this gap and provide comprehensive information on Cotton Breeding in India in one compact volume. Main features of this book are briefly presented as follows:

This book contains 12 chapters which cover topics like brief history and problems, evolution of cultivated cottons, cotton species and races, use of male sterility, breeding methods, cotton breeding in north, central and south zones transgenic breeding, cotton seed as a source of edible oil, DUS testing and Indian seed legislations. The information contained in this book has been gathered from various published sources and Internet websites. Attempts have been made to provide latest information even then some valuable information might have been missed.

The lists of Indian cotton varieties and hybrids, varieties and hybrids under seed production chain, area, production and yield of cotton in India, list of transgenic *Bt.* cotton hybrids have been appended. The Glossary of important terms used in cotton breeding has been provided at the end for ready reference. Important References have been listed at the end which would be useful in gathering further details.

The cooperation extended and the patience shown by my wife Mrs. Jaswanti Singh during completion of this manuscript is highly appreciable. Hope this Volume would be useful to the students, researchers, plant breeders and seed producers engaged with cotton crop. Constructive suggestions of readers are welcome for further improvement of this book.

Phundan Singh

Sanjeev Singh

Contents

Seed Oil Content; Oil Estimation Techniques; Practical Achievements; Future Thrusts; Summary; Questions.

Chapter 1

History and Problems

INTRODUCTION

Cotton is an important fibre yielding crop which plays key role in the world's economy. Cotton refers to those species of the genus *Gossypium* which possess spinnable seed coat fibres. In other words, cotton possesses lint that can be spun in to fine yarn. There are some other plant species which bear seed coat fibre, but the fibre is not spinnable. Such species include silk cotton or kapok [*Ceiba pentendra*] and Madar [*Calotropis* spp.]. Some facts about cotton are given below.

(i) Cotton belongs to the genus *Gossypium* of the family Malvaceae.

(ii) Cotton is basically a perennial crop but forced to grow as annual.

(iii) Cotton is basically a self pollinated species with average cross pollination of 5-6 per cent.

(iv) Cotton bears seed coat fibres that can be spun in to fine yarn.

(v) Cotton fibres have twisting property which makes it suitable for spinning.

(vi) Cotton is also known as white gold.

(vii) Cotton retains its reputation as queen of the fibres.

(viii) Cotton is a multipurpose crop. It is a fibre, oil and protein yielding crop.

(ix) In India, now transgenic *Bt.* cotton is predominant.

SPECIES OF COTTON

In the genus *Gossypium* about 50 species have been identified so far. Out of these four species, *viz.*, *Gossypium hirsutum*, *Gossypium barbadense*, *G. arboreum* and *G. herbaceum* are cultivated and rest are wild or uncultivated. The first two species are tetraploid [2n =52] and last two species are diploid [2n =26]. Tetraploid cottons are also known as New World cottons and diploid cottons are also known as Old World cottons or Asiatic cottons. India is the only country where all the four cultivated species are commercially planted. The *G. hirsutum* is also known as upland cotton or American cotton. *Gossypium barbadense* is also known as Egyptian cotton or Peruvian cotton or Tanguish cotton or Sea Island cotton.

BRIEF HISTORY OF COTTON BREEDING

In India cotton crop is grown from time immemorial. The cotton improvement work started with human civilization when man started selecting superior plants of cotton for making clothes. In India, systematic cotton improvement work started in 1904 when Department of Agriculture was established in different states. In India, history of cotton improvement can broadly be divided in to three parts, *viz.*, (1) varietal era, (2) hybrid era, and (3) transgenic cotton era. A brief description of these three eras is presented as follows:

1. Varietal Era

The development of cotton varieties started with human civilization and continued till 1970. Initially land races were grown. Systemic cotton improvement work started since 1904 with the establishment of Agriculture department in different states. In 1947 [before partition], about 97 per cent of the cotton area was under diploid cottons that is about 70 per cent under *G. arboreum* and 27 per cent under *G. herbaceum*. Later on, upland cotton gradually replaced diploid cottons on vast area because of its high yield potential and good fibre properties. The main points related to varietal era are highlighted as follows:

   (i) This era started with human civilization.

  (ii) Diploid cotton cultivars dominated prior to independence.

 (iii) Tetraploid cottons got acclimatized to Indian conditions.

 (iv) After independence, upland cotton replaced diploid cottons gradually on vast areas.

  (v) There was gradual improvement in yield, quality and maturity duration.

 (vi) About 200 cotton varieties were developed in all the four cultivated species for different cotton growing states.

2. Hybrid Era

The hybrid cotton era started since 1970 when world's first cotton hybrid [H 4 or Shankar 4] was released from Main Cotton Research, Surat by Dr C. T. Patel for commercial cultivation in Gujarat. Later on this hybrid spread to many other states such as Maharashtra, Madhya Pradesh, Karnataka and Andhra Pradesh by virtue of its wider adaptability and high yield potential. Subsequently, several hybrids including *intra-hirsutum* hybrids, *hirsutum x barbadense* hybrids and *arboreum* x herbaceum hybrids were released for different states. The main points related to hybrid era are highlighted as follows:

  (i) This era started with the release of H 4 in 1970.

 (ii) Hybrids gradually replaced varieties of both *hirsutum* and diploid cottons on vast areas.

(iii) Three types of hybrids, *viz.*, *intra-hirsutum, hirsutum x barbadense* and diploid hybrids were developed for commercial cultivation in different states.

(iv) Both conventional and male sterility based hybrids were developed.

(v) There was gradual improvement in yield, quality and disease resistance.

(vi) About 60 hybrids were developed by public sector and about 120 hybrids by private seed companies.

3. *Bt.* Cotton Era

The *Bt.* cotton era started since 2002 when government of India permitted cultivation of three transgenic cotton hybrids in India. Initially *Bt.* hybrids with single gene [Bollgard I] were developed and later on *Bt.* hybrids with two genes [Bollgard II] were released. Gradually *Bt.* hybrids spread under cultivation replacing conventional cotton hybrids as well as varieties on vast areas. The main points related to *Bt.* cotton era are highlighted as follows:

(i) This era started in 2002 with the release of three *Bt.* hybrids for cultivation in India.

(ii) *Bt.* cotton hybrids gradually replaced non *Bt.* hybrids and cotton varieties on vast areas.

(iii) There was appreciable improvement in yield.

(iv) There was effective control of bollworms.

(v) More than 1200 *Bt.* hybrids have been released by 35 private seed companies.

(vi) During 2009-2010, the area under *Bt.* hybrids was about 85 per cent of total cotton area.

(vii) Both bollgard I and bollgard II hybrids are under cultivation.

(viii) Both *intra-hirsutum and hirsutum x barbadence* hybrids have been released.

COTTON GROWING PROBLEMS

In India, there are several problems which are associated with cotton cultivation. These problems are related to crop improvement, crop production, crop protection and cotton technology. The major problems are listed below:

1. Related to Crop Improvement

(i) Multiplicity of *Bt.* cotton hybrids.

(ii) Cultivation of unidentified/unreleased genotypes.

(iii) Inadequate availability of certified seed.

(iv) Sale of spurious *Bt.* cotton seed.

(v) Sale of F_2 seed of *Bt.* cotton.

(vi) High cost of hybrid seed of *Bt.* cotton.

(vii) Problems of hybrid seed production especially in North zone.

2. Related to Crop Production

    (i) Delayed sowing.

   (ii) Use of low doses of fertilizers.

  (iii) Imbalanced use of fertilizers.

  (iv) Poor plant population due to high seedling mortality.

   (v) Soil salinity in some areas.

  (vi) Excessive soil moisture/water stagnation in some areas.

 (vii) Improper spacing especially plant to plant.

(viii) Excessive plant growth especially in the northern zone.

  (ix) Untimely rains in some areas.

3. Related to Crop Protection

    (i) High incidence of bollworms in all the three zones.

   (ii) High incidence of Cotton Leaf Curl Virus [ClCuV] in northern zone.

  (iii) Spurious insecticides and use of low doses of insecticides.

  (iv) Indiscriminate use of insecticides.

   (v) Poor knowledge about pesticide application – it includes choice, dose, time of application and method of application of insecticides.

  (vi) Inadequate availability of bio-agents, light traps and pheromone traps.

 (vii) Emergence of new insects such as mealy bug and mirid bug.

4. Cotton Technology

There are certain requirements of fibre technology which are rarely fulfilled. Important technological requirements are as follows:

    (i) Low neps and motes contents specially in inter-specific hybrids.

   (ii) Low short fibre contents.

  (iii) High fibre elongation.

  (iv) High fibre strength and adequate fibre fineness.

   (v) Low seed coat fragments.

  (vi) High fibre maturity, *etc.*

MAJOR ACHIEVEMENTS

In India, remarkable achievements have been made in cotton improvement. Several high yielding varieties and hybrids were released from various Cotton Research Centres for cultivation in different cotton growing areas. Improvement has been made in fibre quality, disease resistance, insect resistance, adaptability, maturity duration. However, *Bt.* cotton has gradually replaced conventional cotton varieties and hybrids [non-Bt. cotton] on vast area. Presently, about 85 per cent of the total cotton area is occupied by *Bt.* cotton in India. Till 2009, about 521 *Bt.* cotton hybrids and one *Bt.* variety [BN Bt.] were released and approved by the central government for commercial cultivation in India. All the 521 *Bt.* cotton hybrids have been developed by Private Seed Companies and one variety [BN Bt.] jointly by Central Institute for Cotton Research, Nagpur and University of Agricultural Sciences, Dharwad.

CONTRIBUTION OF SOME COTTON BREEDERS

The significant contribution of some renowned Indian and foreign cotton breeders is briefly presented as follows:

[A] Indian Plant Breeders

C.T. Patel

He was a famous Cotton Breeder at the Main Cotton Research Station of Gujarat Agricultural Universiy [now Navsary Agricultural University, Navsary, Gujarat]. He developed the world's first cotton hybrid [H4 or Shankar 4] for commercial cultivation in 1972 for which he was awarded Padamshree. His pioneer work laid the foundation hybrid cotton in India. For his remarkable contribution in cotton, he is known as the father of hybrid cotton and Gujarat is known as the home of the hybrid cotton.

V. Santhanam

He is an eminent Cotton Breeder who has developed several high yielding varieties of upland cotton and two varieties of Egyptian cotton [Sujatha and Suvin]. He was the first Project Coordinator of cotton and later on worked in Food and Agricultural Organization of United Nations as cotton consultant.

B.H. Katarki

He was a famous Cotton Breeder at the University of Agricultural Sciences, Dharwad. He developed World's first inter-specific hybrid between *G. hirsutum* and *Gossypium barbadense* named as Varalaxmi in 1972 for commercial cultivation in Karnataka State. This hybrid has long staple and is capable of spinning 80 counts. Initially, it was released for irrigated areas of karnatak, but later on spread to Tamil Nadu, Andhra Pradesh and Maharashtra. His work encouraged cotton breeders and later on several interspecific tetraploid hybrids were released.

K. Srinivasan

He was a famous Cotton Breeder at the Central Institute for Cotton Research, Regional Station, Coimbatore. He developed the World's first male sterility based hybrid in upland cotton in 1978 under the name Suguna for cultivation in Tamil Nadu State.

N.P. Mehta

He was Senior Research Scientist [Cotton] at the Main Cotton Research Station, Surat of Navsari Agricultural University, Gujarat. Dr. N. P. Mehta and his team developed world's first interspecific diploid hybrid DH 7 between *G. herbaceum* and *G. arboreum* in 1985. Three years later in 1988 another inter-specific diploid hybrid DHI with long staple was released for cultivation in Gujarat State.

V.N. Shroff

He was a famous Cotton Breeder at Jawaharlal Nehru Krishi Vishwa Vidyalaya Campus Indore. Dr. Shroff and his team developed the first *intra-hirsutum* hybrid JKHy 1 in 1976 for cultivation in Madhya Pradesh. Later on this hybrid became very popular in Andhra Pradesh. He also worked on male sterility in cotton.

M.A. Tayyab

He was Senior Cotton Breeder at Punjabrao Deshmukh Krishi Vidyapeeth, Akola. He developed World's first cytoplasmic genic male sterility based hybrid *i.e.* PKV Hy 3 in upland cotton in 1993 for cultivation in Vidarbha region of Maharashtra. He put the hybrid research work on sound footings. Later on new sources of genetic male sterility in *G. aboreum* and new source of cytoplasmic genic male sterility in *G. aridum* were identified at PDKV, Akola by Dr. L.D. Meshram and his colleagues.

T.H. Singh

He was a Senior Cotton Breeder at Punjab Agricultural University, Ludhiana. Dr. T. H. Singh and his team developed the first *intra-hirsutum* hybrid [Fateh] in 1994 for cultivation in Punjab State. Simultaneously, an intra-arboreum hybrid [LDH 11] was developed from PAU, Ludhiana by D. B. S. Sandhu for Punjab State.

Munshi Singh

He was a Senior Cotton Breeder at Indian Agricultural Research Institute, New Delhi. Dr. Munshi Singh and his team developed several high yielding varieties of upland cotton for north zone [Rashmi, Pusa Ageti, Pusa 31, Pusa 8-6 and Pusa 595 B]. His tean also developed advanced breeding material in upland cotton with very high fibre strength.

H.G. Singh

He was a famous Cotton Breeder. He developed high yielding varieties of cotton [Pramukh, Shamli and Lohit]when he was Assistant Economic Botanist at Bulandshahar. Later he became Head, Department of Plant Breeding and Genetics at Chandra Shekhar Azad University of Agriculture and Technology, Kanpur and Vice Chancellor at G. B. Pant University of Agriculture and Technology, Pantnagar.

Phundan Singh

He is an eminent Cotton Breeder. He transferred locule retention ability in *Gossypium arboreum* from the race *cernuum* and developed several locule retentive genotypes of *G. arboreum* cotton. Some of such lines have been registered with National Bureau of Plant Genetic Resources [New Delhi} as unique germplasm lines. He has created ample literature in the field of Plant Breeding and Genetics in general and cotton breeding in particular. He was Director of the Central Institute for Cotton Research, Nagpur.

B.P.S. Lather

He was Chief Scientist Cotton at Chaudhary Charan Singh Haryana Agricultural University, Hisar. Dr. Lather and his team developed the first *intra-hirsutum* cotton hybrid [Dhanlaxmi] in 1994 for cultivation in Haryana State.

D.P. Singh

He was a Senior Cotton Breeder at Chaudhary Charan Singh Haryana Agricultural University, Hisar. Dr. D.P. Singh and his team identified male sterility in *G. arboreum* and developed first genetic male sterility based *intra-arboreum* hybrid [AAH 1] in 1999 for cultivation in Haryana State.

R.P.S. Bhardwaj

He was a Senior Cotton Breeder at Regional Agricultural Station, Sriganganagar of Rajasthan Agricultural University, Bikaner. Dr. Bhardwaj and his team developed the first *intra-hirsutum* hybrid [Maru Vikas] in 1994 for cultivation in Rajasthan State.

[B] Foreign Cotton Breeders

J.B. Hutchinson

He was a famous British Botanist. He worked at the Institute of Plant Industry, Indore [Madhya Pradesh] and attempted inter-specific hybridization between *G. hirsutum* and *G. tomentosum*. Later on some varieties [Badnawar 1, Khandwa 1 and Khandwa 2] were released from the interspecific derivatives for cultivation in Madhya Pradesh. Hutchinson, Silow and Stephens classified species of *Gossypium* in to different sections. Hutchinson also presented description of different races of *G. hirsutum*, *Gossypium barbadense* and *G. herbaceum*.

R.A. Silow

He was an English Botanist. He in association with J.B. Hutchinson classified the genus *Gossypium* into different sections. Silow also presented detailed characteristics of different races of *G. arboreum*.

S.C Harland

He was an English Geneticist. Harland in 1940 proposed that *Gossypium hirsutum* has evolved from a cross between cultivated diploid species *G. arboreum* and wild American species *G. thurberi*. The cross between these species took place in nature and chromosome doubling probably by cosmic radiation resulted in development of fertile tetraploid species. Later on Phillips reported that upland cotton has evolved from a cross between *G. africanum* [2n= 26, linted] and *G. raimondii* [2n =26,lintless]. Now latter view is widely accepted.

L.S. Bird

He was an American Cotton Breeder. He developed the concept of Multiple Adversities Resistance [MAR] and developed several genotypes combining such characters.

FUTURE THRUSTS

In future cotton demand will increase due to population pressure. To keep pace with the increasing cotton demands, the future cotton breeding efforts have to be directed towards following thrust areas:

1. Development of cytoplasmic-genic male sterility (CGMS) based *Bt.* hybrids for irrigated and rain-fed conditions.

2. Development of short duration (165 days) tetraploid or diploid *Bt.* hybrids with four tones of seed cotton yield/hectare for northern irrigated conditions.

3. Development of *Bt.* hybrids and varieties suitable for machine picking.

4. Development of short duration, short stature and compact *Bt.* cultivars and hybrids in upland cotton to achieve quantum jump in the productivity by adopting closer spacing.

5. Development of *Bt.* hybrids and cultivars resistant to moisture stress conditions suitable for rain dependent cotton cultivation.

6. In north zone, there is an increasing incidence of leaf curl virus in upland cotton (*G. hirsutum*). Hence, there is a need to develop varieties and hybrids of upland cotton resistant to leaf curl virus for this zone.

7. Development of *Bt.* hybrids and cultivars suitable for late sowing.

8. Development of *Bt.* hybrids and cultivars with wide adaptability.

9. Development of *Bt.* hybrids and cultivars with high fibre strength suitable for high speed (jet and rotor) spinning. Development of such cultivars is also necessary to compete in the global market.

10. Development of *Bt.* varieties in *G.barbadense* better than Suvin.

11. Development of neps and motes free intra- *barbadense* hybrids for irrigated areas of Tamil Nadu.

12. Development of *Bt.* varieties and hybrids with high ginning outturn [more than 40 per cent] both in diploid and tetraploid cottons.

SUMMARY

Cotton refers to those species of the genus *Gossypium* which possess spinnable seed coat fibres. In other words, cotton possesses lint that can be spun in to fine yarn. There are some other plant species which bear seed coat fibre, but the fibre is not spinnable. Such species include silk cotton or kapok [*Ceiba pentendra*] and Madar [*Calotropis* spp.].

In the genus *Gossypium* about 50 species have been identified so far. Out of these four species, *viz.*, *Gossypium hirsutum*, *Gossypium barbadense*, *G. arboreum* and *G. herbaceum* are cultivated and rest are wild or uncultivated. The first two species are tetraploid [2n =52] and last two species are diploid [2n =26]. Tetraploid cottons are also known as New World cottons and diploid cottons are also known as Old World cottons or Asiatic cottons.

The cotton improvement work started with human civilization when man started selecting superior plants of cotton for making clothes. In India, systematic cotton improvement work started in 1904 when Department of Agriculture was established in different states. In India, history of cotton improvement can broadly be divided in to three parts, *viz.*, (1) varietal era, (2) hybrid era, and (3) transgenic cotton era. A brief description of these three eras has been presented.

In India, cotton growing belt is divided into three zones *viz.*, North zone, Central zone and South zone. Northern cotton growing zone consists of Punjab, Haryana, Rajasthan and Western Uttar Pradesh, where cotton is grown entirely under irrigation in sandy loam soils. Central zone comprises of Madhya Pradesh, Maharashtra and Gujarat. Predominant area is under black soil, (vertisols), which is subjected to runoff, erosion, soil and nutrient losses. Cotton is grown as a mono-crop or as an intercrop. Southern zone includes Andhra Pradesh, Telangana, Karnataka, and Tamil Nadu. Cotton cultivation is done both under irrigated and rainfed conditions. Soils of this zone are both black and red and poor in fertility. In this region *intra-hirsutum* and *hirsutum x barbadense Bt.* hybrids are predominant and some area is under Egyptian cotton and diploid cottons. The area is well known for growing long and extra long staple *G. barbadense* cottons. Cotton is grown in south as sole crop or as intercrop with onion, chilli, cowpea, maize, *etc.* Cotton- rice rotation is also followed in this region.

In India, there are several problems which are associated with cotton cultivation. The problems related to crop improvement, crop production, crop protection and cotton technology have been discussed.

In India, remarkable achievements have been made in cotton improvement. Improvement has been made in fibre quality, disease resistance, insect resistance, adaptability, maturity duration. However, *Bt.* cotton has gradually replaced conventional cotton varieties and hybrids [non-*Bt.* cotton] on vast area. Presently, about 85 per cent of the total cotton area is occupied by *Bt.* cotton in India. Till 2009, about 521 *Bt.* cotton hybrids and one *Bt.* variety [BN *Bt.*] were released and approved by the central government for commercial cultivation in India. All the 521 *Bt.* cotton hybrids have been developed by Private Seed Companies and one variety [BN *Bt.*] jointly by Central Institute for Cotton Research, Nagpur and University of Agricultural Sciences, Dharwad. Future thrusts of cotton improvement have been presented.

QUESTIONS

1. Define cotton and describe briefly important characteristics of cotton.

2. Describe briefly the history of cotton breeding in India.

3. Explain in detail cotton growing problems in India.

4. Discuss briefly the future thrusts of cotton breeding in India.

5. **Describe the contribution of following cotton breeders.**
   (i)   C.T. Patel   (ii) V. Santhanam
   (iii) B.H. Katarki   (iv) M.A. Tayyab

6. **Discuss the contribution of following cotton breeders.**
   (i)   J.B. Hutchinson   (ii)   R.A. Silow
   (iii) S.C. Harland   (iv) L.S. Bird

7. **Write short notes on the following:**
   (i)   White gold   (ii)   Old World cotton
   (iii) New World cotton   (iv) Upland cotton

Chapter 2

Evolution of Cultivated Cottons

INTRODUCTION

Cotton refers to those species of the genus *Gossypium* which possess spinnable seed coat fibres. In other words, cotton possesses lint that can be spun in to fine yarn. There are some other plant species which bear seed coat fibre, but the fibre is not spinnable. Such species include silk cotton or kapok [*Ceiba pentendra*] and Madar [*Calotropis* spp.]. Cotton is inherently a semi-xerophytic perennial crop. However, it is being grown as an annual/seasonal crop.

CENTRES OF ORIGIN

There are four species of cotton, *viz.*, *Gossypium hirsutum* [upland cotton], *Gossypium barbadense* [Sea Island cotton], *G. arboreum* [tree cotton] and *G. herbaceum* [Levant cotton]. Centres of origin refer to geographical areas where crop plants have originated. Centres of origin are of two types, *viz.*, primary centres of origin and secondary centres of origin. A primary centre of origin is the original home of a crop species and secondary centres of origin are regions with high diversity which have developed as a result of subsequent spread of a crop. In cotton, the centre of origin of each cultivated species is different.

1. *Gossypium herbaceum*

It is believed that cultivated species *Gossypium herbaceum* has probably differentiated from African linted but wild species *G. africanum* in South Africa and later on it spread to other countries. However, some researchers are of the opinion that The *Gossypium herbaceum* originated probably in Pakistan [Sindh] after differentiation from *Gossypium africanum*. From their centre of origin these two diploid species spread to other countries.

2. *Gossypium arboreum*

It is believed that cultivated species *Gossypium arboreum* has differentiated from African linted but wild species *G. africanum* in India and later on it spread to other countries. It is believed that *Gossypium arboreum* originated in India probably in Gujarat after differentiation from *Gossypium africanum*.

3. *Gossypium barbadense*

It is believed that the Sea Island cotton originated first and other species afterwards. The Sea Island cotton originated in South America probably in Peru from a cross between two diploid species [2n=26], *viz.*, *Gossypium africanum* and *Gossypium raimondii*. The F1 was sterile which became fertile over a long time by chromosome doubling through cosmic radiations. Thus the centre of origin of *Gossypium barbadense* is Peru in South America.

4. *Gossypium hirsutum*

Central America is considered as the centre of origin of upland cotton [*Gossypium hirsutum* L.]. It is believed that the material of Sea Island cotton reached Mexico through travelers and traders. Over a long time after differentiation it gave birth to the upland cotton. Thus Mexico in Central America is the centre of origin of upland cotton. From Peru and Mexico, these two tetraploid species spread to other parts of the world.

GENETIC ORIGIN

(i) Diploid Cottons

It is believed that diploid species of cotton originated first and tetraplois species later on form a cross between two diploid species. The wild linted species *Gossypium africanum* is considered as the progenitor of all cultivated species of cotton. *Gossypium africanum* reached undivided India from South Africa through travelers, traders and explorers and after differentiation gave birth to two diploid cultivated species, *viz.*, *Gossypium arboreum* and *Gossypium herbaceum*.

(ii) Tetraploid Cottons

There are two species of tetraploid cultivated cotton, *viz.*, *Gossypium hirsutum* and *Gossypium barbadense*. The genome constitution of both these species is AD. It is believed that A genome of tetraploid cotton has come from Asian cultivated species and D genome from American lintless wild diploid species (Skovsted, 1934). Skovsted (1934) first pointed out that the New World tetraploid consisted of A and D sub-genomes. Later on origin of tetraploid cotton was explained by other workers. There are two views about the genetic origin of tetraploid cotton as discussed below:

(1) Beasley and Harland View

In 1940, Beasley and Harland independently observed that A genome of upland cotton has come from Asian cultivated diploid species *Gossypium arboreum* and D genome from American wild lintless diploid species *G. thurberi*. The cross between these two species took place in nature followed by chromosomal doubling. This is represented as follows:

Parents	*Gossypium arboreum* x	*Gossypium thurberi*
Genome	AA [2n=26, large]	DD [2n =26, small
F1		AD [Sterile]
Chromosome doubling		AA DD Fertile – Like G. hirsutum
		[4n+ 52, 26 large and 26 small]

Figure 2.1: Probable Origin of *G. hirsutum* as Proposed Independently by Beasley (1940) and Harland (1940).

(2) Phillips (1963) View

He reported that A genome of upland cotton has come from African linted wild diploid species *Gossypium africanum* and D genome from American wild lintless diploid species *G. raymondii*. The cross between these two species took place in nature followed by chromosomal doubling. This is represented as follows:

Parents	*Gossypium africanum* x	*Gossypium raymondii*
Genome	AA [2n=26, large]	DD [2n =26, small
F1		AD [Sterile]
Chromosome doubling		AA DD Fertile – Like G. *hirsutum*
		[4n+ 52, 26 large and 26 small]

Figure 2-2: Probable Origin of *G. hirsutum* as Proposed Phillips (1963).

Cytological, biochemical [electrophoretic studies] and molecular investigations have clearly indicated that A genome of *G. africanum* is closer to A genome of tetraploid cotton than that of *G. arboreum*. Similarly, D genome of *G. raymondii* is closer to D genome of tetraploid cotton than that of *G. thurberi*. Thus *G. africanum* and *G. raymondii* are progenitors of tetraploid cotton. The second view is widely accepted.

Hutchinson, Silow and Stephens (1947) presented classification of the genus Gossypium on the basis of cytological, genetic, geographical and archeological evidences. They divided all the cultivated and wild species of cotton in to eight sections, *viz., Sturtiana, Erioxyla, Klotzschiana, Thurberana, Anomala, Stockiana, Herbacea* and *Hirsuta*. Later on Phillips added one more section *i.e.* Longicalyciana.

SUMMARY

Cotton refers to those species of the genus *Gossypium* which possess spinnable seed coat fibres. In other words, cotton possesses lint that can be spun in to fine yarn. There are four species of cotton, *viz., Gossypium hirsutum* [upland cotton],

Gossypium barbadense [Sea Island cotton], *G. arboreum* [tree cotton] and *G. herbaceum* [Levant cotton].

It is believed that cultivated species *Gossypium herbaceum* has probably differentiated from African linted but wild species *G. africanum* in South Africa and later on it spread to other countries. It is believed that *Gossypium arboreum* originated in India probably in Gujarat after differentiation from *Gossypium africanum*.

It is believed that cultivated tetraploid species have originated from a cross between Africal wild linted species *Gossypium africanum* and American wild lintless species *G. raymondii*. The cross between these two species took place in nature followed by chromosomal doubling.

It is believed that the Sea Island cotton originated first and other species afterwards. The Sea Island cotton originated in South America probably in Peru from a cross between two diploid species[2n=26], *viz.*, *Gossypium africanum* and *Gossypium raymondii*.

It is believed that the material of Sea Island cotton reached Mexico through travelers and traders. Over a long time after differentiation it gave birth to the upland cotton. Thus Mexico in Central America is the centre of origin of upland cotton.

QUESTIONS

1. Define cotton and describe briefly centres of origin of cultivated cottons.

2. Describe briefly centres of origin of upland and Sea Island cottons.

3. Explain briefly genetic origin of upland cotton and Tanguish cottons.

4. Discuss briefly the genetic origin of upland cotton proposed by Beasley.

5. Describe the genetic origin of upland cotton proposed by Phillips.

6. Discuss the contribution of following cotton breeders.

    (i)   Beasley   (ii) Phillips

    (iii) S.C. Harland   (iv) L.S. Bird

7. Write short notes on the following:

    (i)   Upland cotton   (ii) Sea Island cotton

    (iii) *G. herbaceum* cotton   (iv) *G. arboreum* cotton

Chapter 3

Species and Races

INTRODUCTION

In the genus *gossypium* about 50 species have been identified so far. The species of cotton are divided in to two groups, *viz.*, cultivated, and wild. These are discussed as follows:

CULTIVATED SPECIES

There are four cultivated species,*viz.*, *Gossypium hirsutum* L., *Gossypium barbadense*, *G. arboreum* L. and *G. herbaceum* and rest are wild or uncultivated. Cultivated species are divided in to two groups, *viz.*, diploid cottons and tetraploid cottons. Important features of four cultivated species of cotton are presented as follows:

Diploid Cottons

There are two species of diploid [2n=26] cultivated cottons, *viz.*, *Gossypium herbaceum* and *Gossypium arboreum*. These are also known as Old World cottons. Since these species are mainly grown in the Asian region, they are also known as Asiatic cottons.

(i) *Gossypium arboreum*

This species is also known as tree cotton and Indian cotton because India is considered as the centre of origin of this species. In this species, bracts are more or less triangular and closely invest bud and flower. Bracts have 4-5 teeth at the apex. Flowers have red spot at the base of the petal. Bolls are tapering and profusely pitted with prominent oil glands. At maturity, bolls open widely. There are six races of this species, *viz.*, *Bengalense, Burmanicum, Cernuum, Indicum, Sinense* and *Sudanense*. In India varieties of three races, *viz.*, *Bengalense, Cernuum* and *Indicum* are commercially cultivated.

(ii) *Gossypium herbaceum*

This species is also known as lavent cotton. In this species, bracts flare widely from the base and have 6-8 teeth. Flowers have red spot at the base of the petal. Bolls are round and rarely with prominent shoulders. Bolls are smooth or with few shallow pits and few oil glands.At maturity, bolls open slightly. There are five races of this species, *viz., Acerifolium, Africanum, Kuljianum, Persicum* and *Wightianum*. Cultvated Varieties of this species belong to the race Wightianum. Differences between *G. arboreum* and *G. herbaceum* are presented in Table 3.1.

TABLE 3.1: Differences between *G. arboreum* and *G. herbaceum* Cottons

Sl.No.	Particulars	G. arboreum	G. herbaceum
1	Other name	Tree cotton	Levant cotton
2	Centre of origin	India	Sindh, Afganistan
3	Bracteoles type	closed	Flaring
4	Bracteole size	Longer than broad	Broader than long
5	Bracteole teeth	3 to 4	6 to 8
6	Bolls	Tapering	Round
7	Genome	A2	A1
8.	Races	Six	Five
9.	Global area covered (per cent)	1	Traces

Tetraploid Cottons

This group includes two species, *viz., Gossypium hirsutum* and *Gossypium barbadense*. These species are also known as New World species and generally possess good quality of lint.

(i) *Gossypium hirsutum*

This species is also known as upland cotton or American cotton. In this species, flowers do not have red spot at the base of the petal. The staminal column is short. The anthers are loosely arranged on the staminal column. Anther filaments are larger in the upper region than in the lower region. The boll surface is usually glabrous. There are seven races of this species, *viz., Latifolium, marie-galante, Morrilli, Palmeri, Punctatum, Richmondi* and *Yucatenance*. All cultivated varieties of this species belong to the race *Latifolium*.

(ii) *Gossypium barbadense*

This species is also known as Sea Island cotton or Peruvian cotton or Tanguish cotton or Egyptian cotton or quality cotton. In this species, flowers have red spot at the base of the petal. Bracts are very large which cover the flower bud completely. The staminal column is long on which anthers are compactly arranged. The anther filaments are of same length. Bolls are large and deeply pitted with black oil glands. This species has only one race *i.e. Brasiliense*. Differences between *G. hirsutum* and *Gossypium barbadense* are presented in Table 3.2.

TABLE 3.2: Differences between *G. hirsutum* and *Gossypium barbadense* Cottons

Sl.No.	Particulars	G. hirsutum	G. barbadense
1	Staminal column	Short	Long
2	Arrangement of anthers	Loose	Compact
3	Anthers filaments	Longer above than below	Of same length
4	Capsule surface	Smooth	Pitted
5	Leaf lobes	Often broadly triangular	Often long tapering
6	Seeds	Usually fuzzy	Usually naked
7	Genome	AD1	AD2
8	Centre of origin	Mexico	Peru
9	Other names	American cotton and upland cotton	Peruvian, Sea Island cotton, Tanguish cotton, Egyptian cotton
10	Stem surface	Usually hairy	Usually smooth
11	Races	Seven	One
12	Global area covered (per cent)	89	10

Bracts are united at the base in diploid cottons and free in tetraploid cottons. Diploid cottons have some degree of resistance to sucking pests and drought and salinity. Tetraploid cottond have high yielding capacity and good fibre quality. India is the only country where all the four cultivated species are commercially planted.

WILD SPECIES

There are about 46 wild species of cotton. Out of these, three species,*viz.*, *Gossypium tomentosum*, *Gossypium mustelinum* and *Gossypium darwinii* are tetraploid and rest are diploid. A list of wild and cultivated species along with genome designation and centre of origin is presented in the Table 3.3.

TABLE 3.3: List of wild and cultivated species of the genus Gossypium

Sl.No.	Name of Species	Genome	Centre of Origin
A	**Diploid Species**		
1	*Gossypium africanum*	A	Africa
2	*Gossypium herbaceum*	A1	Afganistan
3	*Gossypium arboreum*	A2	India, Burma, China
4	*Gossypium anomalum*	B1	Africa
5	*Gossypium triphyllum*	B2	Africa
6	*Gossypium capitis-viridis*	B3	Cape verde
7	*Gossypium sturtianum*	C1	Cape Verde
8	*Gossypium robinsonii*	C2	Australia
9	*Gossypium thurberi*	D1	America
10	*Gossypium amourianum*	D2-1	America
11	*Gossypium harknessii*	D2-2	America

Sl.No.	Name of Species	Genome	Centre of Origin
12	*Gossypium klotzchianum*	D3-k	America
13	*Gossypium davidsonii*	D3-d	America
14	*Gossypium aridum*	D4	America
15	*Gossypium raimondii*	D5	America
16	*Gossypium gossypioides*	D6	America
17	*Gossypium lobatum*	D7	America
18	*Gossypium trilobum*	D8	America
19	*Gossypium laxum*	D9	America
20	*Gossypium turneri*	D10	Australia
21	*Gossypium schwendimanii*	D11	Australia
22	*Gossypium stocksii*	E1	Arabia
23	*Gossypium somlense*	E2	Arabia
24	*Gossypium areysianum*	E3	Arabia
25	*Gossypium incanum*	E4	Arabia
26	*Gossypium benaderense*	E	Australia
27	*Gossypium bricchettii*	E	Australia
28	*Gossypium vollesenii*	E	Australia
29	*Gossypium trifucatum*	E ?	Australia
30	*Gossypium longicalyx*	F1	Africa
31	*Gossypium bickii*	G1	Australia
32	*Gossypium australe*	G	Australia
33	*Gossypium nelsonii*	G	Australia
34	*Gossypium anapoides*	K	-
35	*Gossypium costulatum*	K	Australia
36	*Gossypium cunninghamii*	K	Australia
37	*Gossypium exiguum*	K	Australia
38	*Gossypium enthyle*	K	Australia
39	*Gossypium londonderriense*	K	Australia
40	*Gossypium merchantii*	K	Australia
41	*Gossypium nobile*	K	Australia
42	*Gossypium pilosum*	K	Australia
43	*Gossypium populofolium*	K	Australia
44	*Gossypium pulchellum*	K	Australia
45	*Gossypium rotundifolium*	K	Australia
B	**Tetraploid Species**		
46	*Gossypium hirsutum*	AD1	America
47	*Gossypium barbadense*	AD2	America
48	*Gossypium tomentosum*	AD3	Hawai
49	*Gossypium mustelinum*	AD	America
50	*Gossypium darwinii*	AD	America

USEFUL TRAITS

Wild species of cotton possess several desirable traits which can be transferred to cultivated cotton. Some desirable traits include fibre length, fibre strength, fibre fineness, fibre elongation, high ginning percentage, resistance to sucking pests and various diseases.

TABLE 3.4: Useful Traits found in different Wild Species

Sl.No.	Useful Traits	Wild Species
A	**Insect Resistance**	
1	Bollworms (tolerance)	*G. thurberi, G. anomalum, G. raymondii, G. armorium* and *G. somalense*
2	Jassids	*G. anomalum, G. armourianum, G. raymondii* and *G. tomentosum*
3	Whitefly	*G. armourianum*
4	Mites	*G. anomalum*
5	Aphids	*G. davidsonii*
B	**Disease Resistance**	
6	Bacterial Blight	*G. anomalum, G. armourianum, G. raymondii*
7	Fusarium Wilt	*G. sturtianum, G. harknessii* and *G. thurberi*
8	Verticillium Wilt	*G. harknessii*, Maxicanum race of *hirsutum*
C	**Fibre Quality**	
9	Fibre length	*G. anomalum, G. stocksii, G. raymondii, G. areysianum, G. longicalyx*
10	Fibre strength	*G. stocksii, G. areysianum, G. Thurber, G. anomalum, G. sturtianum, G. raymondii* and *G. longicalyx*
11	Fibre elongation	*G. stocksii, G. areysianum, G. Thurber, G. anomalum, G. sturtianum, G. raymondii* and *G. longicalyx*
12	Fibre Fineness	*G. longicalyx, G. anomalum* and *G. raymondii*
13	High ginning	*G. austral*
14	Lint yield	*G. anomalum, G. sturtianum, G. austral, G. stocksii* and *G. areysianum*
D	**Other traits**	
15	Drought Resistance	*G. tomentosum, G. stocksii, G. darwinii, G. areysianum, G. anomalum, G. austral, G. harknessii, G. aridum,* and *G. raymondii*
16	Frost Resistance	*G. thurberi*
17	Male sterility [cytoplasmic]	*G. harknessii, G. trilobum* and *G. aridum*
18	Delayed morpho-genesis	*G. austral* and *G. bickii*

RACES OF CULTIVATED COTTONS

There are seven races of *Gossypium hirsutum* [*Punctatum, Palmeri, Marie-galante, Morilli, Latifolium, Richmondi* and *Yukatenance*], one of *Gossypium barbadense* [*brasiliense*], six of *Gossypium arboreum* [*Bengalense, Burmanicum, Cernuum, Indicum, Sinense* and *Soudanense*] and four of *Gossypium herbaceum* [*Acerifolium, Kuljjianum,*

Persicum and *Wightianum*]. All these races have some useful traits which can be utilized in developing superior cotton cultivars.

TABLE 3.5: Races of Cultivated Species and Useful Traits found in them

Sl.No.	Species and Races	Useful Traits found
A	**Gossypium hirsutum**	
	Punctatum	Resistance to drought, jassids, bacterial blight and Verticillium wilt; fibre fineness, fibre strength and less gossypol.
	Palmeri	Resistance to sucking pests.
	Marie-galante	Resistance to stem weevil.
	Morilli	Resistance to jassids
	Latifolium	Not known
	Richmondi	Not known
	Yukatenance	Not known
B	**Gossypium barbadense**	
	Brasiliense	Resistance to stem weevil.
C	**Gossypium arboreum**	
	Bengalense	Resistance to bacterial blight and high ginning.
	Burmanicum	Fibre strength and fineness.
	Cernuum	Big boll, locule retentivity, high ginning and high absorbancy.
	Indicum	High fibre length and fineness.
	Sinense	Earliness, fibre fineness and strength
	Soudanense	Drought resistance and wide adaptation.
D	**Gossypium herbaceum**	
	Acerifolium	Drought resistance
	Kuljianum	Drought resistance
	Persicum	Drought resistance
	Wightianum	Drought resistance

Thus useful traits found in races of cultivated species include, resistance to drought, jassids, bacterial blight, stem weevil, sucking pests, *Verticillium* wilt; fibre quality traits [fibre length, fibre fineness,high ginning], earliness, high absorbance, *etc.*

SUMMARY

In the genus Gossypium about 50 species have been identified so far. The species of cotton are divided in to two groups, *viz.*, cultivated, and wild. There are four cultivated species,*viz.*, *Gossypium hirsutum*, *Gossypium barbadense*, *G. arboreum* and *G. herbaceum* and rest are wild or uncultivated. The first two species are tetraploid [2n=52] and last two are diploid [2n=26]. The list of all species [cultivated and wild] is presented depicting genome and distribution.

Wild species of cotton possess several desirable traits which can be transferred to cultivated cotton. Some desirable traits include fibre length, fibre strength, fibre fineness, fibre elongation, high ginning percentage, resistance to sucking pests and various diseases.

There are seven races of *Gossypium hirsutum* [*Punctatum, Palmeri,Marie-galante, Morilli, Latifolium, Richmondi* and *Yukatenance*], one of *Gossypium barbadense* [brasiliense], six of *Gossypium arboreum* [*Bengalense, Burmanicum, Cernuum, Indicum, Sinense* and *Soudanense*] and four of *Gossypium herbaceum* [*Acerifolium, Kuljjianum, Persicum* and *Wightianum*]. All these races have some useful traits which can be utilized in developing superior cotton cultivaris.

QUESTIONS

1. Define cultivated and wild cotton and describe briefly their important characteristics.

2. Describe briefly the important features of upland cotton.

3. What is Egyptian cotton? Describe important features of the same.

4. Explain briefly important features of diploid cottons.

5. Discuss briefly the future thrusts of cotton breeding in India.

6. Give a brief comparison of Upland cotton and Sea Island cotton.

7. Explain in brief important differences between tree cotton and lavent cotton

8. Discuss the contribution of following cotton breeders.

9. Give the list of traits of breeding value found in wild species of cotton.

10. List various races of cultivated cottons and describe in brief traits of breeding value found in these races.

11. Describe briefly useful traits found in wild species and races of cultivated cottons.

Chapter 4

Male Sterility

INTRODUCTION

In flowering plants, male sterility refer to a condition in which either pollen is absent or non functional. Use of male sterility helps in reducing the cost of hybrid seed by eliminating the labour oriented process of hand emasculation. Hence, use of male sterility in developing hybrids became essential.

ADVANTAGES OF MALE STERILITY

In cotton, there are two methods of hybrid seed production, *viz.,* (a) hand emasculation and pollination method; also called conventional method, and (b) male sterility method. The conventional method has several drawbacks such as (1) high cost of hybrid seed, (2) low seed setting, (3) low crossing efficiency, (4) high shedding of crossed bolls, (5) chances of selfing, (6) more immature seeds, (7) check on F_2 seeds, (8) less production of hybrid seed per unit area, (9) more labour requirement, *etc*. On the other hand, the use of male sterility in hybrid seed production has several advantages. The main advantages of male sterility method of hybrid seed production in relation to conventional method are discussed below:

1. Cheaper Hybrid Seed

Use of male sterility helps in reducing the cost of hybrid seed production due to elimination of emasculation process. The reduction in cost of hybrid seed production has been reported upto 40 per cent. The cost of hybrid seed production is about Rs. 150/- per kg in male sterility hybrids against Rs. 350/- per kg in conventional hybrids. For production of one kg seed, one labour is required per day against 3 laborers per day in conventional method.

2. Higher Seed Setting

The seed setting is poor in conventional method of hybrid seed production due to mutilation of ovary during emasculation. The seed setting is high when hybrid seed is produced using male sterile line, because there is no mutilation of ovary in this method.

3. Lesser Immature Seeds

The proportion of immature seeds is lesser when the hybrid seed is produced using male sterility as compared to hand emasculation and pollination method of hybrid seed production.

4. Lesser Crossed Bolls Shedding

The shedding of crossed bolls is lesser when male sterility is used for hybrid seed production as compared to conventional method of hybrid seed production. This is because of mutilation of ovary in the latter.

5. Free from Selfed Seed

There is no chance of selfed seed production when male sterility is used for hybrid seed production. In conventional method, there are chances of selfing.

6. High Crossing Efficiency

The efficiency of crossing is very high when male sterility is used for hybrid seed production. One labourer can cross 600-900 flowers per day against 200-300 flowers per day in conventional method.

7. Check on Use of F_2 Seed

In conventional hybrids, possibilities of selling F_2 seeds are there. Such possibilities are not there in hybrid seed produced using CGMS, because the whole population in F_2 will be sterile.

8. Higher Hybrid Seed Production

Through use of male sterility, more quantity of hybrid seed can be produced than conventional method. This can be attributed to three factors, *viz.*, (1) higher seed setting, (2) higher crossing efficiency, and (3) and higher planting ratio of female and male parents *i.e.* 4:1 or 3:1 (2:2 in conventional method).

9. Rapid Spread of Hybrids

The reduction in cost of hybrid seed production and production of higher quantity of seed leads to rapid spread of hybrids under cultivation. When the seed is cheaper, even marginal and small farmers would like to grow hybrids.

OBJECTIVES OF USING MALE STERILITY

There are three main objectives of using male sterility in developing cotton hybrids as follows:

1. Reduction in Cost of Hybrid Seed

The hybrid seed of conventional cotton hybrids is very high (ranging from Rs. 700/- per kg to Rs. 1000/- per kg) which cannot be afforded by small and marginal farmers. High cost of hybrid seed acts as a barrier in spread of hybrids under cultivation. Use of male sterility in hybrid seed production will help in making the hybrid seed available to the farmers at cheaper rate.

2. Diversification of Male Sterility Sources

In upland cotton, at present, we have one source of genetic male sterility *i.e.* Gregg male sterility and two sources of cytoplasmic-genic male sterility *i.e.*, *G. harknessii* and *G. aridum*. To avoid danger of uniformity, there is need to diversify sources of male sterility. The sources of male sterility can be diversified in two ways, *viz.*, (1) by interspecific crosses involving wild species, and (2) by induced mutagenesis.

3. Conversion of Elite Lines into Male Sterility

There is need to convert parents of commercially cultivated hybrids and elite germplasm lines into GMS, CMS and R lines to develop heterotic pool for future use.

MALE STERILITY GENES IDENTIFIED

In cotton, 18 loci for male sterility have been identified so far. Of these, 13 loci have been identified in upland cotton (*Gossypium hirsutum* L.), three in Egyptian cotton (*Gossypium barbadense* L.) and two in diploid cotton (*Gossypium arboreum* L.). Male sterility is governed by both recessive and dominant genes (Table 4.1). However, cases of recessive male sterility are higher than dominant male sterility. Recessive male sterility is governed by single gene as well as double genes. But dominant male sterility is governed by single gene. Dominant loci for male sterility have been identified in upland as well as Sea Island cottons. In *G. arboreum*, each of the two male sterility loci identified so far is governed by single recessive gene.

MALE STERILITY GENES IN USE

In upland cotton, 13 loci for male sterility have been identified. However, only Gregg male sterile line which is governed by double recessive genes (ms_5ms_6) is used for developing commercial hybrids. The Gregg male sterility, though governed by two recessive genes, is functional for one recessive gene only. The genetic constitution of male sterile line is ms_5ms_5/ms_6ms_6 and that of maintainer line is either Ms_5ms_5/ms_6ms_6 or ms_5ms_5/Ms_6ms_6. Thus, the maintainer line is heterozygous at one locus only. In *G. arboreum* cotton, two genes *viz.*, ams_1 and $ar.ms$ are being used in developing commercial hybrids. There are four main drawbacks of genic male sterility (GMS) as given in Table 4.1.

1. Roguing

In GMS, 50 per cent of the population which is male fertile has to be rogued out every year which makes the hybrid seed production costly due to production of less quantity of hybrid seed.

2. Less Stable

The GMS, especially in diploid cotton is temperature sensitive. Sterile plants become fertile especially under low temperature *i.e.* below 16°C.

TABLE 4.1: Male Sterility Gene Identified in Cotton

Sl.No.	Species	Genes Identified	Reference
1.	G. hirsutum	Recessive	
		ms_1	Justus and Leinweber, 1960
		ms_2	Richmond and Kohel, 1961
		ms_3	Justus et al. 1963
		ms_5ms_6	Weaver, 1968
		ms_8ms_9	Rhyne, 1971
		ms_{14}, ms_{15}, ms_{16}	Zhang et al. 1994
		Dominant	
		MS_4	Allison and Fisher, 1964
		MS_7	Weaver and Ashley, 1971
		MS_{10}	Bowman and Weaver, 1979
2.	G. barbadense	Recessive	
		ms_{13}	Percy and Turcotte, 1991
		Dominant	
		MS_{11}	Turcotte and Feaster, 1979
		MS_{12}	Turcotte and Feaster, 1985
3.	G. arboreum	Recessive	
		ams_1	Singh and Kumar, 1993
		ar.ms	Meshram et al. 1994

3. Late Identification

The identification of male sterile and male fertile plants is possible only after anthesis.

4. Chances of Admixture

There are chances of admixture if fertile plants are not rogued out properly.

CYTOPLASMIC GENETIC MALE STERILITY

In cotton, cytoplasmic genic male sterility (CGMS) was first reported by Meyer in 1975 and he also identified a source of fertility restoration. The fertility is restored by a single dominant gene (Rf). The cytoplasmic genic male sterility is highly stable because it is not influenced by environmental factors. In cotton, two practically usable sources of cytoplasmic genic male sterility have been identified. The first source is wild diploid species *i.e. G. harknessii* which was identified in USA. The second source is again a diploid species *i.e. G. aridum* which was identified in India. Both these sources are being used for developing hybrids. Main advantages and drawbacks of CGMS system are as follows:

Advantages

The main advantages of cytoplasmic genic male sterility are summarized below :

(i) Highly Stable

The cytoplasmic genic male sterility is highly stable because it is not affected by environmental factors such as temperature and day length. CMS line will always produce male sterile plants.

(ii) Easy Conversion

The conversion of a genotype into CMS is faster. Any genotype can be converted into CMS line by making 5-6 continuous backcrosses.

(iii) Easy Maintenance

The maintenance of CMS line requires lesser area than that of GMS line.

(iv) No Roguing

All the plants of CMS line are male sterile. Hence, there is no need of roguing. Thus, more quantity of hybrid seed can be produced using CMS line as compared to GMS line. Moreover, expenditure on roguing is curtailed.

Drawbacks

The main drawbacks of cytoplasmic genic male sterility system are briefly presented below:

(i) Limited Restorers

The female parent of the hybrid has to be converted into CMS and male parent into restorer R line. Moreover, there are limited number of restorer lines which restrict the use of CGMS system.

(ii) Adverse Effects

The cytoplasmic genes controlling the male sterility have sometimes adverse effects on other characters. For example, the *G. harknessii* source of CMS is susceptible to sucking pests. However, the *G. aridum* source of CMS is resistant/tolerant to sucking pests. As a result, now *G. aridum* source is more widely being used than *G. harknessii* source in developing hybrids.

(iii) Low Yielder

The presently available A and R lines are poor yielding. As a result, the yield of CMS based hybrid is 10-15 per cent lower than conventional hybrid developed using the same parents.

(iv) Poor Restoration Capacity of R Lines

The restoration capacity of R lines is poor. Moreover, there is variation in the restoration capacity among different plants of the same R line.

(v) Narrow Genetic Base

The presently available CMS and R lines have narrow genetic base resulting in poor performance.

MAJOR ACHIEVEMENTS

The work on use of male sterility in developing cotton hybrids was intensified since 1999 when a National Agricultural Technology Project (NATP) was launched on hybrid cotton in India. There are three major achievements of research work done under this project which are as follows:

1. Release of Commercial Hybrids

The first genetic male sterility based intra-*hirsutum* hybrid *i.e.* Suguna was released in 1978. Recently, two types of hybrids, *viz.*, (i) Intra-*hirsutum,* and (ii) intra-*arboreum* have been released for commercial cultivation using male sterility. Three intra-*hirsutum* hybrids *viz.*, PKVHY 3, PKVHY 4 and PKVHY 5 have been developed using cytoplasmic genic male sterility. Three intra-*arboreum* hybrids *viz.*, AAH1, AKDH7 and G. Cot. MDH 11 (Tables 4.2 and 4.3) have been developed using genetic male sterility. In upland cotton, two hybrids (PKVHY 3 and PKVHY 4) have been developed using *G. harknessii* source of CMS and one hybrid (PKVHY 5) using *aridum* source of CMS. In *G. arboreum,* two hybrids (AAH 1 and G. Cot. MDH 11) have been developed using Hisar source of GMS and one (AKDH 7) using Akola source.

TABLE 4.2: Male Sterility based Hybrids Released in Cotton

Sl.No.	Type of Hybrid	Male Sterility Source Used	Hybrids Released
1.	Intra-*hirsutum* (CMS)	G. harknessii	PKV HY3
		G. harknessii	PKV HY4
		G. aridum	PKV HY5
2.	Intra-*arboreum* (GMS)	Hisar (ams$_1$)	AAH 1
		Hisar (ams$_1$)	G.Cot MDH 11
		Akola (ar.ms)	AKDH 7

The important features and yield performance of these male sterility hybrids are presented in Table 4.3.

2. Identification of New Male Sterility Sources

In India, two new male sterility sources each in upland and *G. arboreum* cottons have been identified since 1993. In upland cotton, wild species *G. aridum* has been identified as new source of cytoplasmic genic male sterility at Akola centre through interspecific hybridisation. In upland cotton, new GMS has been obtained through induced mutation from variety Abadhita (10 kR gamma rays + 0.2 per cent EMS combination). In *G. arboreum,* two male sterility loci (ams$_1$ – as spontaneous mutant of variety DS 5 and ar.ms from *G. anomalum*) have been identified (Table 4.4).

TABLE 4.3: Important Features of Male Sterility Based Hybrids

Sl.No.	Name of Hybrid	Year of Release	Yield (q/ha)	GOT (per cent)	MFL (mm)	Spinning Counts	Area for which Released for
\multicolumn{8}{c}{(i) Intra-*hirsutum* hybrids (CMS)}
1.	Suguna	1978	30	35	25	40	Tamil Nadu
2.	PKV HY3	1993	15R	36	25	40	Vidarbha
3.	PKV HY4	1996	20R	35	30	50	Vidarbha
4.	PKV HY5	2000	15R	35	26	40	Vidarbha
\multicolumn{8}{c}{(ii) Intra-*arboreum* hybrids (GMS)}
5.	AAH 1	1999	24	38	16	< 10	Haryana
6.	AKDH 7	2000	12R	38	24	30	Vidarbha
7.	G.Cot MDH 11	2002	20	36	23	27	Gujarat

TABLE 4.4: New Male Sterility Sources Identified in Cotton in India

Sl.No.	Species	Type of Male Sterility	Sl.No.	Source	How Obtained/Method Used
1.	G. hirsutum	CMS	1.	G. aridum	Interspecific hybridisation
2.	G. hirsutum	GMS	2.	Abadhita	Induced mutation
3.	G. arboreum	GMS	3.	ams$_1$	Spontaneous mutant (DS5)
			4.	G. anomalum	Interspecific hybridisation

3. Conversion into GMS, CMS and R Lines

In upland cotton, large number of genotypes have been converted into CMS, GMS and R lines. The Gregg source for conversion into GMS and *G. harknessii* and *G. aridum* sources for conversion into CMS and R lines were used (Table 4.5). In diploid cotton, Hisar and Akola sources (ams$_1$ and ar.ms) were used for conversion of elite lines into GMS.

TABLE 4.5: Cotton Genotypes Converted into CMS, R and GMS Lines

Sl.No.	Species	Converted into	Sl.No.	Source Used	Lines Converted
1.	Tetraploid	CMS	1.	G. harknessii	537
			2.	G. aridum	67
		R	3.	G. harknessii	231
			4.	G. aridum	806
		GMS	5.	Gregg	75
2.	Diploid	GMS	6.	Hisar and Akola	130

FUTURE THRUSTS

Some cotton hybrids have been released using male sterility, new sources of male sterility have been identified and several elite lines have been converted into

CMS, R and GMS lines. The future research work on the use of male sterility need to be directed towards following thrust areas:

1. Development of CGMS in Diploid Cotton

In diploid cotton, only genetic male sterility is available in which 50 per cent of the population of female parent has to be rogued out for hybrid seed production. Hence, there is need to develop cytoplasmic genic male sterility system in diploid cotton.

2. Development of Temperature Insensitive GMS

The genetic male sterility available at present in diploid cotton is temperature sensitive. The male sterile plants become fertile at temperature below 16°C. Hence, there is need to develop temperature insensitive genetic male sterility in diploid cotton.

3. Improvement in Restoration Capacity of R Lines

In upland cotton, the restoration capacity of R lines is poor. Moreover, there is variation in the restoration capacity within the line. Hence, efforts have to be made to improve restoration capacity of R lines. This can be achieved in two ways *i.e.* (i) by inter-mating of plants of same R line having high restoration capacity, and (ii) by inter-mating of two R lines having high restoration capacity.

4. Development of Interspecific Hybrids

Intraspecific male sterility based hybrids have been developed in upland and *G. arboreum* cottons. There is ample scope to develop male sterility based interspecific hybrids at tetraploid and diploid levels.

5. Use of Marker in GMS Lines

In genetic male sterility, 50 per cent of the population has to be removed for hybrid seed production. Moreover, the sterile plants can be identified after anthesis only. There is need to develop GMS lines with marker character so that fertile plants can be rogued out during seedling stage.

6. Use of Biotechnology

In rapeseed (Canola), male sterility has been developed through biotechnology. In cotton also biotechnological approach may be used to develop male sterility.

SUMMARY

Use of male sterility helps in reducing the cost of hybrid seed by eliminating the labour oriented process of hand emasculation. Hence, efforts were made to develop male sterility based hybrids so that hybrid seed could be provided to the farmers at cheaper rate. The first genetic male sterility based intra-*hirsutum* hybrid *i.e.* Suguna was released in 1978. Subsequently, two types of hybrids, *viz.*, (i) Intra-*hirsutum*, and (ii) intra-*arboreum* were released for commercial cultivation using male sterility. Three intra-*hirsutum* hybrids *viz.*, PKVHY 3, PKVHY 4 and PKVHY 5 were developed using cytoplasmic genic male sterility. Three intra-*arboreum* hybrids *viz.,* AAH1, AKDH7 and G.Cot MDH 11 (Tables 4.2 and 4.3) were developed using genetic male sterility. In upland cotton, two hybrids (PKVHY 3 and PKVHY 4) were developed using *G. harknessii* source of CMS and one hybrid (PKVHY 5) using *G. aridum* source of CMS. In *G. arboreum,* two hybrids (AAH 1 and G. Cot. MDH 11) were developed using Hisar source of GMS and one (AKDH 7) using Akola source. The future thrust areas for developing male sterility based hybrids have been indicated.

QUESTIONS

1. Define male sterility and describe various advantages of male sterility.

2. List male sterility genes reported in cotton. Describe commonly used gene in practical cotton breeding.

3. What is genetic male sterility? Describe its merits and demerits.

4. Explain briefly merits and demerits of cytoplasmic genic male sterility.

5. Discuss briefly objectives of using male sterility in cotton.

6. Explain in brief practical achievements of using male sterility in cotton.

7. Discuss the thrusts of male sterility in cotton.

8. Give a brief account of cotton hybrids released using male sterility in cotton.

9. Write short notes on the following:

    (i)   Cytoplasmic genic male sterility   (ii) Genetic male sterility

    (iii) Demerits of GMS   (iv) Demerits of CGMS

Chapter 5

Breeding Methods

INTRODUCTION

Plant breeding techniques that are used for genetic improvement of cotton refer to cotton improvement procedures. Cotton is basically a self pollinated crop. The crop improvement procedures commonly used in cotton are of two types, *viz.*, general breeding methods and special breeding methods as outlined below:

1. General Breeding Method

These methods include plant introduction, pure line selection, mass selection, pedigree method, bulk breeding method and backcross method.

2. Special Breeding Methods

These methods include heterosis breeding, mutation breeding, wide crossing, polyploidy breeding and transgenic breeding.

Detailed account of these methods is beyond the scope of the present discussion for which readers may refer Cotton Breeding by Dr. Phundan Singh. A brief account of cotton improvement through application of these breeding methods is presented as follows:

PLANT INTRODUCTION

Transposition of crop plants from the place of their cultivation to such areas where they were never grown earlier is called plant introduction. It is an ancient and rapid method of crop improvement. The main drawbacks of plant introduction are that some times new weeds, insects and diseases may get entry in to new area or country. The introduced plant material is used in three ways, *viz.*, (i) directly as a variety, (ii) as a variety after selection, and (iii) as a parent in hybridization for developing new varieties or hybrids. Plant introduction has been used in cotton improvement. A summary of uses of introduced material in cotton improvement is presented in Table 5.1.

TABLE 5.1: Utilization of Plant Introductions

Sl.No.	Material Introduced	Introduced from	Used as
1	Sea Island Andrews	West Indies	Directly released as a variety in Tamil Nadu.
2	Egyptian Karnak	Egypt	Released as Sujata after selection.
3	Short branch types	Russia	Released after selection as Variety PRS 72.
4	American Nectariless	America	Resulted in development of World's first cotton Hybrid 4 after crossing with Gujarat 67.
5	SB 289E	Russia	Resulted in development of Hybrid Varalaxmi after crossing with Laxmi.
6	SB 1085	Russia	Resulted in development of Hybrid CBS 156 after crossing with Acala glandless from America.

PURE LINE SELECTION

Pure line refers to the homogeneous progeny of a sel pollinated homozygous plant obtained by selfing. Pure line selection is an old method of cotton improvement. Pure line selection has been widely used for cotton improvement and several improved varieties have been released. Some examples of practical achievements by pure line selection are presented species wise in Table 5.2.

TABLE 5.2: Practical Achievements of Pure Line Selection

Sl.No.	Cotton Species	Examples of Varieties Developed by Pure Line Selection
1	*Gossypium hirsutum*	MCU 5 VT, CO2 and LSS.
2	*G. barbadense*	Sujata.
3	*G. arboreum*	Cocanada 2, Gaorani 22, Gaorani 46 and Lohit.
4	*G. herbaceum*	Western 1 and Selection 69.

MASS SELECTION

Mass selection has been widely used in cotton and several improved varieties have been developed. Examples of some cotton varieties developed in upland and *arboreum* cotton are presented in the Table 5.3.

TABLE 5.3: Cotton Varieties Developed by Mass Selection

Sl.No.	Cotton Species	Examples of Varieties Developed by Mass Selection
1	*Gossypium hirsutum*	F 414, Bikaneri Narma, H 777, Pramukh, SRT 1, Narmada, L 147, PRS 72, Mysore Vijay, *etc.*
2	*Gossypium arboreum*	G 27, HD 11, LD 133, gaorani 6, AK 277, Cocanada White, Saraswati, *etc.*

PEDIGREE METHOD

Pedigree method has been extensively used in cotton improvement and several improved varieties have been developed. Examples of some cotton varieties developed by this method are presented species wise in Table 5.4.

TABLE 5.4: Examples of Varieties Developed by Pedigree Method

Sl.No.	Cotton Species	Examples of Varieties Developed by Pedigree Method
1	Gossypium hirsutum	J 34, J 205, LH 372, LH 900, LH 1134, LH 1556, F 505, F 286, F 846, F 1054, F 1378, HS 45, HS 6, H 1098, H 974, RST 9, Ganganagar Ageti, RS 875, Vikas, Khandwa 3,G. Cot. 14, DHY 286, PKV 081, Rajat, Nagnath, Sharda, Abhadita, JK 119, Sahana, MCU 6, MCU 8, MCU 9, MCU 11, LRA 5166, LRK 516, Surabhi, CNH 36, Arogya, *etc.*
2	G. barbadense	Suvin
3	G. arboreum	Shamli, Maljari, AK 235, AKH 4, Srisailam, Sanjay, Mahanandi, Jyoti, K 7, K 8, K 9, K 10, K 11, G. Cot.15, Namdeo, AKA 8401, DS 5, HD 107, LD 230, LD 327, LD 491, *etc.*
4	G. herbaceum	Jaydhar, Vijalpa, G. Cot. 11, G. Cot. 13, G. Cot. 17, G. Cot. 21, G. Cot. 23, *etc.*

BULK BREEDING METHOD

This method has been rarely applied for genetic improvement of cotton especially in India.

BACK CROSS METHOD

This method is used specially for transfer of disease resistance and male sterility from one genotype to another. By this method some varieties have been developed in *Gossypium herbaceum* such as V 797, Vijalpa, Kalyan, *etc.*

HETEROSIS BREEDING

This method has been extensively used for genetic improvement of cotton and several hybrids have been developed for commercial cultivation in India. Both types of hybrids, *viz.*, non-Bt. hybrids and *Bt.* cotton hybrids have been developed by this method in India. Some examples of the achievements of heterosis breeding are presented in the Table 5.5.

TABLE 5.5: Practical Achievements of Heterosis Breeding

Sl.No.	Type of Hybrid	Examples of Hybrids Developed
1	Intra-hirsutum	H 4, H 6, H 8, H 10, JKHy 1, JKHy 2, NHH 44, Savita, Fateh, Dhanlaxmi, Maru Vikas, Omshankar, Surya, *etc.*
2	Hirsutum x barbadense	Varalaxmi, DCH 32, NHB 12, HB 224, DHB 105, TCHM 213, Sruthi, *etc.*
3	Intra-arboreum	LDH 11, AAH 1, AKDH 7, Raj DH 16, *etc.*
4	Herbaceum x arboreum	DH 7, DH 9, DDH 2, Pha 46.
5	Male sterility based hybrids	PKV Hy 3, PKV Hy 4, PKV Hy 5, Suguna, MECH 4, MECH 11, Ankur 9, Ankur 15, AKDH 7, AAH 1, G. Cot. MDH 11, CICR Hy 2.
5	Bt. Cotton hybrids	In all 521 for cultivation in India.

MUTATION BREEDING

In cotton, mutation breeding has been used by several workers and some improved varieties have been developed. The summary of achievements of mutation breeding in cotton is presented in the Table 5.6.

TABLE 5.6: Varieties of Cotton Developed by Mutation Breeding

Sl.No.	Cotton Species	Varieties Developed	How Developed ?
1	*Gossypium hirsutum*	Indore 2	Through X-ray irradiation of Malwa upland 4.
		MCU 7	Through X-ray irradiation of L 1143.
		Rashmi	Through gamma irradiation of MCU 5.
		Pusa Agetii	Through X-ray irradiation of Stoneville 213.
		MCU 10	Through X-ray irradiation of MCU 4.
		LK 861	This is spontaneous mutant of variety Krishna.
		CNH 120 MB	Through EMS treatment of Variety SRT 1.
2.	*Gossypium arboreum*	DS 1	Through gamma irradiation of G 27.
3.	*Gossypium herbaceum*	DB 3-12	This is spontaneous mutant of variety Western 1.

DISTANT HYBRIDIZATION

Crossing between different species of the same genus or between two different genera of the same family is called distant hybridization or wide crossing. Thus, distant hybridization is of two types, *viz.*, inter-specific hybridization and inter-generic hybridization. In cotton, inter-specific hybridization has been used to considerable extent, whereas inter-generic hybridization has been used to very limited extent. Distant hybridization is associated with problems of cross incompatibility, hybrid in-viability and hybrid sterility. Inter-specific hybridization is used when desirable gene is not found within the species. Interspecific hybridization has been used for improvement in quality, disease resistance, insect resistance, drought resistance, identification of male sterility *etc*. Several varieties and hybrids of cotton have been developed through inter-specific hybridization [Table 5.7].

Table 5. 7: Varieties and Hybrids Developed through Interspecific hybridization

Sl.No.	Cotton Species/Hybrids	Varieties/Hybrids Developed
A	**Varieties**	
1	*Gossypium hirsutum*	Badnawar 1, Khandwa 1, Khandwa 2, Deviraj (170 Co2), Gujarat 67, Devitej (130 Co2 M), MCU 2, MCU 5, PKV o81, Rajat, Arogya, *etc.*
2	*Gossypium arboreum*	AKA 8401
B	**Hybrids**	
1.	*hirsutum x barbadense*	Varalaxmi, DCH 32, DHB 105, NHB 12, TCHB 213, HB 224, Sruthi, *etc.*
2	*Herbaceum x arboreum*	DH 7, DH 9, DDH 2, Pha 46, *etc.*

POLYPLOIDY BREEDING

In cotton, polyploidy breeding has been rarely used. It has been used mainly in combination with distant hybridization for inter-specific gene transfer.

TRANSGENIC BREEDING

It refers to development of cotton varieties/hybrids through application of biotechnology. In India, transgenic breeding has been extensively used for genetic improvement of cotton.In India 521 *Bt.* cotton hybrids have been released till 2009 by 35 different seed companies. Out of these, 164 *Bt.* hybrids have been released for north zone, 296 for central zone and 294 for south zone. One Variety *i.e.* BN *Bt.* has also been developed for cultivation in all the three zones.

Besides above methods, there are some plant breeding approaches which are used for population improvement. These techniques include recurrent selection, biparental mating and diallel selective mating. For detailed account of these approaches, readers may refer Cotton Breeding by Phundan Singh.

SUMMARY

Plant breeding techniques that are used for genetic improvement of cotton refer to cotton improvement procedures. Cotton is basically a self pollinated crop. The crop improvement procedures commonly used in cotton are of two types, *viz.*, general breeding methods and special breeding methods. General Breeding methods include plant introduction, pure line selection, mass selection, pedigree method, bulk breeding method and backcross method. Special Breeding methods include heterosis breeding, mutation breeding, wide crossing, polyploidy breeding and transgenic breeding. A brief account of cotton cultivars developed through application of these breeding methods has been presented.

QUESTIONS

1. List various breeding methods used in cotton. Describe any one of them in detail.

2. Define plant introduction and describe its role in cotton breeding.

3. What is pure line selection? Describe its role in cotton breeding.

4. Define pedigree breeding. Describe its role in cotton breeding with examples.

5. Define heterosis. Describe the role of heterosis in cotton breeding.

6. Explain briefly the role of mutation breeding in cotton with examples.

7. What is wide crossing? Discuss its role in cotton improvement with examples.

8. Define transgenic breeding. Describe its role in cotton improvement.

9. Describe briefly cotton varieties developed through mutation breeding and heterosis breeding.

10. **Write short notes on the following:**

   (i)   Plant Introduction   (ii) Pure line selection

   (iii) Distant hybridization   (iv) Transgenic breeding

Chapter 6

Breeding in North Zone
[Punjab, Haryana and Rajasthan]

INTRODUCTION

In India, cotton growing belt is divided into three zones *viz.*, North zone, Central zone and South zone. Northern cotton growing zone consists of Punjab, Haryana, Rajasthan and Western Uttar Pradesh, where cotton is grown entirely under irrigation in sandy loam soils. Upland cotton varieties and hybrids hybrids are predominant in this region and some area is under *arboreum*. Cotton-wheat is the predominant cropping system. This chapter deals with cotton breeding in north zone. Main features of cotton cultivation in north zone are as follows:

1. The cotton crop is cultivated under irrigated conditions.
2. The soils are alluvial with plain topography.
3. The yield level of Punjab is the highest in the North zone. The varieties of *G. arboreum* cultivated in Punjab possess short staple.
4. The cotton crop is grown in the kharif season and sowing is done from 15th April to May end.
5. This is a multiple cropping zone and cotton-wheat is a common crop rotation.

MAJOR PROBLEMS

Important problems of north zone related to cotton cultivation are listed below:

1. The temperature is very high during seedling stage which leads to high seedling mortality resulting in poor plant stand.

2. In some areas, there is problem of soil salinity and water logging which leads to reduction in yield.

3. There is problem of leaf curl virus in upland cotton varieties and hybrids which causes considerable yield loss.

4. The cotton crop is grown under irrigated conditions and soil fertility in north zone is also high. Both these factors lead to excessive vegetative plant growth resulting in low harvest index and difficulty in picking.

5. The peak flowering period coincides with peak rainfall period. Hence, there is difficulty in hybrid seed production. This zone has to depend on central or south zone for hybrid seed production.

6. The labour is very costly which also works as a constraint in hybrid seed production.

7. The temperature goes below 15°C after 15th November and as a result boll opening is ceased after 15th of November.

1. PUNJAB STATE

In Punjab, cotton is grown in the entire state. However, the major cotton growing districts of Punjab are Bathinda, Faridkot, Ferozpur and Sangrur.

Species Composition

Prior to independence, almost entire area was under *G. arboreum* race *Bengalense*. The *Gossypium hirsutum* was first introduced in 1850. Presently, only two species of cotton, *viz., G. hirsutum* and *G. arboreum* are grown. The Maximum area is occupied by Intra-hirsutum *Bt.* hybrids followed by *G. arboreum*. All varieties and hybrids of upland cotton have been replaced by *Bt.* cotton hybrids.

Breeding Centres

Prior to independence, cotton improvement work was being carried out at Lyallpur [now in Pakistan] and Abohar. Now, the cotton breeding work is carried out at two research centres, *viz.,* Faridkot and Ludhiana under Punjab Agricultural University, Ludhiana. Faridkot is the main research centre and Ludhiana is the sub-centre. Earlier, Ludhiana was the main centre and Faridkot was the sub-centre.

Prior to independence, plant introduction, pure line selection and mass selection were applied for selection of superior varieties both in upland cotton and diploid cotton. The hybridization work was first initiated in 1930 in *G. arboreum* at Lyallpur. Now all breeding techniques, *viz.,* introduction, mass selection, pure line selection, hybridization, mutation and transgenic breeding are used for genetic improvement of cotton. Cotton improvement work was intensified after the establishment of Punjab Agricultural University at Ludhiana.

Progress of Breeding

Remarkable progress has been made in cotton improvement in Punjab state. Important achievements of cotton breeding in Punjab since 1975 onwards are presented as follows:

(i) Development of short duration varieties (LH 886 and PAU 626).

(ii) Release of intra-*hirsutum* and intra-*arboreum* hybrids (Fateh and LDH 11).

(iii) Development of long staple *hirsutum* varieties (LH 1134, LH 1556).

(iv) Development of leaf-curl virus resistant cultivars and hybrids of upland cotton (LHH 144).

Practical Achievements

Important varieties of upland cotton, *G. arboreum* cotton and hybrids developed from Punjab are presented in Appendix 1. The varieties which were under cultivation prior to release of *Bt.* Hybrids included LH 900, LH 1566, F 846 and F 1378 in *G. hirsutum*; and LD 327, LD 491 and LD 694 in *G. arboreum*. Two intra-*hirsutum* hybrids *viz.*, Fateh and LHH 144 and two intra-*arboreum* hybrid (LDH 11 and PAU 626) were released. Hybrid LHH 144 has okra leaf and is resistant to leaf curl virus. The *arboreum* cotton cultivars are comparable to *hirsutum* cultivars in yield potential. Several *Bt.* Cotton hybrids were released by private seed companies.

2. HARYANA STATE

Haryana is the second largest grower of cotton in the northern zone. In Haryana cotton is grown in the entire state. However, the major cotton growing districts of Haryana include Hisar, Sirsa and Jind.

Species Composition

In Haryana region of greater Punjab [Himachal Pradesh, Punjab and Haryana], prior to independence, almost entire area was under *G. arboreum* race *Bengalense*. The *Gossypium hirsutum* was first introduced in 1850. Presently, only two species of cotton, *viz.*, *G. hirsutum* and *G. arboreum* are grown. The Maximum area is occupied by intra-*hirsutum Bt.* hybrids followed by *G. arboreum*. All varieties and hybrids of upland cotton have been replaced by *Bt.* cotton hybrids.

Breeding Centres

Before 1947 [Independence], the cotton improvement work was being carried out for the entire greater Punjab at Lyallpur [now in Pakistan] and Abohar. Now, the cotton breeding work for Haryana State is carried out at two research centres, *viz.*, Hisar and Sirsa under Chaudhary Charan Singh Haryana Agricultural University, Hisar. The main research centre is at the university headquarters *i.e.* Hisar and Sirsa is the sub-centre. Before partition of Greater Punjab in to three states, *viz.*, Himachal Pradesh, Punjab and Haryana, the cotton breeding work for Haryana was being carried at Hansi. The Regional Station of Central Institute for Cotton Research located at Sirsa also caters to the needs of north zone. The main focus of these breeding centres is to develop productive cultivars of *G. hirsutum* and *G. arboreum* and also intraspecific hybrids of these species that can fit well into the cotton-wheat cropping system of Haryana. The Cotton Section of IARI is concentrating on development of high fibre strength cultivars/hybrids in upland cotton for northern cotton belt of India.

The important cotton breeding procedures applied for selection of superior varieties both in upland cotton and diploid cotton, prior to independence included, plant introduction, pure line selection and mass selection. The hybridization work was first initiated in 1930 in *G. arboreum* at Lyallpur. Now all breeding techniques, *viz.*, plant introduction, pure line selection, mass selection, hybridization, mutation and transgenic breeding are used for genetic improvement of cotton. Cotton improvement work was intensified after the establishment of Haryana Agricultural University at Hisar.

Progress of Breeding

Important achievements of cotton breeding in the Haryana State are presented as follows:

1. Development of short duration varieties.
2. Release of intra-*hirsutum* and intra-*arboreum* hybrids (Dhanlaxmi, Om Shankar, AAH 1, HHH 223, CSHH 198, CICR 2).
3. Identification of genetic male sterility in *G. arboreum*.
4. Development of GMS based diploid hybrids (AAH-1).
5. Development of leaf-curl virus resistant cultivars and hybrids of upland cotton (HHH 223).

Practical Achievements

Important varieties of upland cotton, *G. arboreum* cotton and hybrids developed from Haryana since 1991 are presented in Appendix 1. The varieties which were under cultivation till recently included HS 6, H 974, H 1098, HS 182 and H 1117 in *G. hirsutum;* and DS 5, HD 107 and HD 123 in *G. arboreum*. In Haryana, four hybrids have also been released. Three hybrids are intra-*hirsutum viz.*, Dhanlaxmi, Om Shankar and HHH 223. The last hybrid has been recently released which is resistant to leaf curl virus. One GMS based intra-*arboreum* hybrid *i.e.* AAH 1 has been released. However, now all varieties and hybrids of upland cotton have been replaced by *Bt.* cotton. Varieties of *G. arboreum* are still occupying some area [10 per cent].

3. RAJASTHAN STATE

Rajasthan is an important cotton growing state of northern cotton belt of India. In Rajasthan, cotton crop is grown on 5.82 lakh hectares and the production is about 11.50 lakh bales. The major cotton growing districts are Sriganganagar, Bhilwara and Banswara.

Species Composition

In Rajasthan, only two species of cotton, *viz.*, *G. hirsutum* and *G. arboreum* are grown. The Maximum area is occupied by *Intra-hirsutum Bt.* hybrids followed by *G. hirsutum* varieties and *G. arboreum*. Now, non-*Bt.* hybrids and varieties of upland cotton have been replaced by *Bt.* cotton hybrids on vast area.

Breeding Centres

In Rajasthan, first *G. arboreum* race *bengalense* was introduced from 1931 to 1941 by people of Punjab who migrated to Rajasthan and settled in Ganga canal area. Later on *hirsutum* material from Kanpur and Indore was introduced in Mewar and Sriganganagar districts and grown under tank, well and canal irrigation. The breeding material of both *arboreum* and *hirsutum* cottons was evaluated at Udaipur centre to select superior types. After the establishment of Rajasthan Agricultural University, Bikaner, the Regional Agricultural Research Station was established at Sriganganagar where cotton breeding work is carried out. The main focus is to develop productive cultivars of *G. hirsutum* and *G. arboreum* and also intra-specific hybrids of these species that can fit well into the cotton-wheat cropping system of Rajasthan. Banswara is the sub-centre for Rajasthan state.

The Regional Station of Central Institute for Cotton Research located at Sirsa also caters to the needs of northern zone including Rajasthan state. The main focus of these breeding centres is to develop productive cultivars of *G. hirsutum* and *G. arboreum* and also intra-specific hybrids of these species that can fit well into the cotton-wheat cropping system of Rajasthan. The Cotton Section of IARI carries out breeding for northern zone.

The important cotton improvement procedures applied for selection of superior varieties both in upland cotton and diploid cotton, prior to independence included, plant introduction, pure line selection and mass selection. Now all breeding techniques, *viz.*, plant introduction, pure line selection, mass selection, hybridization, mutation and transgenic breeding are used for genetic improvement of cotton. Cotton improvement work was intensified after the establishment of Rajasthan Agricultural University at Bikaner.

Progress of Breeding

Important achievements of cotton breeding in Rajasthan State are presented as follows:

1. Development of short duration varieties.
2. Release of intra-*hirsutum* and intra-*arboreum* hybrids (Maru Vikas and Raj DH 7).
3. Development of leaf curl virus resistant cultivars and hybrids of upland cotton (LHH 144).

Practical Achievements

Important varieties of upland cotton, *G. arboreum* cotton and hybrids developed in Rajasthan since 1991 are presented in Appendix 1. The currently cultivated varieties include RST 9, RS 875 and RS 810 in *G. hirsutum* and RG 8 and RG 18 in *G. arboreum*. The variety RS 810 has been recommended for cultivation in the entire north zone. Both varieties of *G. arboreum* are also grown in entire north zone. One intra-*hirsutum* hybrid *i.e.* Maruvikas has been released.

FUTURE PROSPECTS

In north zone, significant achievements have been made in cotton breeding in the past. In future cotton breeding work in this zone needs to be directed towards the following thrust areas:

1. All the varieties of *G. arboreum* available in the north zone so far have short staple. There is ample scope to develop *G. arboreum* cultivars with medium and long staple.

2. Most of the currently cultivated varieties and hybrids of upland cotton are susceptible to leaf-curl virus. There is an urgent need to develop leaf curl resistant *Bt.* Hybrids and varieties of upland cotton.

3. The fibre strength of most of the released varieties is poor. There is need to develop *Bt.* varieties and hybrids with high fibre strength suitable for high speed spinning.

4. The ginning percent of cotton varieties especially in upland cotton is low compared to varieties of other countries. There is ample scope to develop *Bt.* varieties and hybrids with high ginning outturn (above 40 per cent) for higher recovery of lint.

5. In some areas, there is problem of soil salinity and water logging. There is need to develop *Bt.* cultivars having capacity to withstand soil salinity and water logging.

6. In order to reduce the cost of *Bt.* cotton hybrids, there is need to develop GMS or CMS based *Bt.* hybrids with high yield potential.

7. In north zone, cotton crop suffers from high temperature during seedling stage. There is urgent need to develop *Bt.* varieties/hybrids having resistance to high temperature during seedling stage.

8. Most of the *Bt.* Hybrids grown have medium staple length. There is ample scope for developing productive *Bt.* varieties and hybrids with high fibre length and strength suitable for export purpose.

9. The cotton picking has become very expensive due to higher labour wages. There is need to develop *Bt.* cotton cultivars/hybrids with short stature and synchronous maturity suitable for machine picking.

SUMMARY

In northern zone, cotton is grown under irrigated conditions and sowing is done from 15th April to 15th May. In Punjab, two species of cotton, *viz.*, *G. hirsutum* and *G. arboreum* are commercially cultivated. Cotton leaf-curl virus is the major problem. Several varieties of upland cotton and arboreum species and few hybrids were released which were till recently under cultivation. Now 164 *Bt.* cotton hybrids have been approved by the government of India for cultivation in north zone. *Bt.* cotton hybrids have replaced upland varieties and hybrids on vast area. Future thrust areas of cotton improvement in this zone have been outlined.

QUESTIONS

1. Explain main features of cotton cultivation in North Zone.

2. Discuss briefly main problems of cotton breeding in North Zone.

3. Describe briefly progress of cotton breeding in North Zone.

4. Give a brief account of practical achievements of cotton breeding in North Zone.

5. Discuss briefly future thrusts of cotton breeding in North Zone.

6. Explain in brief practical achievements of cotton breeding in:

   (i)   Punjab   (ii) Haryana

   (ii) Rajasthan   (iv) Western Uttar Pradesh

7. Give a brief account of cotton hybrids released in North Zone.

8. Explain the role of Central Institute for Cotton Research in breeding varieties for North Zone.

9. Describe the contribution of following cotton breeders:

   (i)   B.P.S. Lather   (ii) Munsi Singh

   (iii) T.H. Singh   (iv) R.P. Bhardwaj

Chapter 7

Breeding in Central Zone
[Madhya Pradesh, Maharashtra and Gujarat]

INTRODUCTION

Central zone comprises of Madhya Pradesh, Maharashtra and Gujarat. Predominant area is under black soil, (vertisols), which is subjected to runoff, erosion, soil and nutrient losses. Cotton is grown as a mono-crop or as an intercrop. In this region also intra-*hirsutum Bt.* hybrids are predominant and some area is under diploid cottons. This chapter deals with cotton breeding in central zone. Main features of cotton breeding in central zone are given below:

1. In this zone, about 70 per cent of the cotton crop is cultivated under rain-fed conditions.

2. The cotton is grown on black cotton soils. The topography is somewhere plain and somewhere undulated.

3. The yield level of seed cotton ranges from 12-15 quintals per ha.

4. Most of the cotton varieties and hybrids grown in this state possess medium and long staple.

5. The cotton crop is grown in the *kharif* season and sowing is done with the onset of monsoon.

MAJOR PROBLEMS

The important problems of central zone related to cotton cultivation are listed below:

1. About 70 per cent cotton crop is grown under rain-fed conditions which often suffers from drought at one or the other stage.

2. More than 50 per cent of cotton growing soils are shallow and have poor water holding capacity.

3. There is lack of varieties and hybrids which can be grown with low inputs.
4. Excessive use of pesticides in some areas resulting in development of resistance in the pests against pesticides.
5. Problem of water logging in some areas is due to improper drainage system.
6. Poor or low productivity due to rain-fed cultivation.
7. Poor fibre quality especially of varieties.
8. Excessive soil moisture during seedling stage.
9. Inadequate availability of good quality seed.
10. There are too many varieties of cotton resulting in contamination.

1. MADHYA PRADESH

There are three major cotton growing zones. Madhya Pradesh comes under central cotton zone of India. Madhya Pradesh covers about 6.8 per cent of total cotton area and contributes 11.6 per cent to the total cotton production.

Growing Regions

In Madhya Pradesh, cotton is cultivated in the entire state. However, the cotton growing area of Madhya Pradesh is divided into four regions, *viz.*, (1) Malwa region, (2) Nimar region, (3) Northern region, and (4) Narmada Valley and Satpura region.

1. Malwa Region

This region includes Badnawar, Ratlam and Rajgarh districts and covers about 1 lakh ha. area. In this region, the cotton is grown as a rainfed crop.

2. Nimar Region

This region is divided into two parts *viz.*, West Nimar and East Nimar. West Nimar includes Khargone, Dhar and Badwani districts and East Nimar consists of Khandwa district. The maximum cotton is grown in the Nimar region which covers about 3.5 lakh ha. In Nimar region, about 50-60 per cent cotton crop is grown as irrigated. For irrigation, the water is lifted from Narmada river.

3. Northern Region

It includes Guna and Gwalior districts and covers about 0.25 lakh ha area. In this region, cotton is grown as a rainfed crop.

4. Narmada Valley and Satpura Region

This region includes Khategaon, Katani, Kannoj, Hoshangabad and Chhindwara districts. This region covers 0.25 lakh ha area and cotton is grown as a rainfed crop.

Species Composition

In this state, two species of cotton *viz.*, *G. hirsutum* and *G. arboreum* are cultivated. Earlier about 45 per cent areas was each under intra-*hirsutum* hybrids and *G. hirsutum* varieties and 10 per cent was under *G. arboreum* varieties. Now *Bt.* cotton hybrids are pre-dominant covering vast cotton area. The remaining area is under *G. hirsutum* and *G. arboreum* varieties. Thus *Bt.* cotton hybrids have replaced *hirsutum* hybrids and varieties on vast area.

Breeding Centres

In Madhya Pradesh, cotton breeding work is carried out at Agricultural Research Station of Jawaharlal Nehru Krishi Vishwa Vidyalaya, Khandwa and Indore. Khandwa is the main research centre and Indore is the sub-centre. The main focus is to develop high yielding varieties of upland and *arboreum* cottons and intra-*hirsutum* hybrids. Earlier Indore was the main cotton breeding centre. Later on, Khandwa was made the main cotton research centre and Indore as sub-centre. Initially, the main cotton in Malwa and Nimar tract was *G. arboreum* race *bengalense*. The Georgian upland cotton was first introduced in 1842 and Cambodian upland in 1912. Pure line selection was made in *G. arboreum* cotton to isolate high yielding and wilt resistant genotypes.

Prior to independence, the Institute of Plant Industry was located at Indore, where famous English Cotton Breeder Sir J.B. Hutchinson worked for several years especially on inter-specific hybridization and generated introgressed material which later on resulted in release of three productive cultivars [Badnawar 1, Khandwa 1 and Khandwa 2].

Initially, plant introduction, pure line selection and mass selection methods were used for cotton improvement. Now all breeding techniques, *viz.*, introduction, mass selection, pure line selection, hybridization, mutation and transgenic breeding are used for genetic improvement of cotton.

Progress of Breeding

The important achievements of cotton breeding in Madhya Pradesh are as given below:

1. Development of insect and disease resistant cultivars [Khandwa 1 and Khandwa 2].
2. Development of short duration cultivars.
3. Development of productive hybrids [JKHy 1 and JKHy 2].
4. Improvement in fibre length and spinning capacity.
5. Improvement in ginning outturn.
6. Development of first coloured linted variety of *G. hirsutum*.

Practical Achievements

In Madhya Pradesh, remarkable progress has been made in cotton breeding after the inception of AICCIP in 1967 (Appendix 1). Varieties of *G. hirsutum*, *G. arboreum* and hybrids developed from this research included Khandwa-2, Khandwa-3, Vikram and JK-4 in *G. hirsutum*; and Maljari, Jawahar Tapti and Sarvottam in *G. arboreum*. Besides varieties, two intra-*hirsutum* hybrids *viz.*, JKHy-1 and JKHy-2 were released which covered about 45 per cent of the total cotton area till recently in the state. Hybrid JKHy-1 has wider adaptability and as a result became very much popular in the state of Andhra Pradesh where it was grown on considerable area.

2. MAHARASHTRA STATE

Maharashtra is the largest cotton growing state in the country. It covers more than 30 per cent of total cotton area and contributes 23 per cent to the production.

Species Composition

In this state, two species of cotton *viz., G. hirsutum* and *G. arboreum* are cultivated, besides hybrids. Mostly intra-*hirsutum* hybrids are grown. Earlier about 60 per cent area was under hybrids, 25 per cent under upland varieties and 15 per cent under *G. arboreum* cultivars. Now maximum cotton area is under *Bt.* cotton hybrids and only 8 per cent under diploid cotton [*G. arboreum*]. Thus *Bt.* cotton hybrids have replaced upland non *Bt.* hybrids and varieties on vast area.

Growing Regions

In Maharashtra, cotton crop is grown in the entire state except Western Maharashtra. However, the cotton growing area of Maharashtra is divided into four major regions, *viz.,* (1) Vidarbha region, (2) Marathwada region, (3) Khandesh region, and (4) Deccan Canal area. These are briefly discussed below:

1. Vidarbha Region

This is the most important cotton growing region of Maharashtra where cotton is cultivated on about 16 lakh hectares. The major cotton growing districts of Vidarbha region are Yavatmal, Amravati, Akola, Buldhana, Wasim, Wardha and Nagpur. In this region, cotton crop is grown under rainfed conditions. Maximum area is covered by intra-*hirsutum* hybrids followed by varieties of upland and <u>*G. arboreum*</u> cottons.

2. Marathwada Region

This is the second largest cotton growing region of Maharashtra. It covers about 8 lakh hectares under cotton. In this region also cotton is cultivated mostly as rainfed crop. The major cotton growing districts include Nanded, Parbhani, Aurangabad and Jalna. In this region, some irrigated cotton (5 per cent) is also grown.

3. Khandesh Region

Khandesh is third largest cotton growing region of Maharashtra. In this region, cotton is cultivated as rainfed crop. In this region, cotton crop is grown on about 2.5 lakh hectares. The main cotton producing districts are Jalgaon, Dhulia and Nandurbar.

4. Deccan Canal Region

In this region, cotton crop is cultivated on a few thousand (about 50-60) hectares under irrigated conditions. In this region, mainly hybrids are grown. Main cotton producing districts are Ahmednagar and Satara.

Breeding Centres

In Maharashtra, cotton breeding work is carried out at the following three main research centres:

1. Cotton Section, Dr. Panjabrao Deshmukh Krishi Vidyapeeth, Akola.
2. Agricultural Research Station, Marathwada Agricultural University, Nanded.
3. Cotton Section, Mahatma Phule Krishi Vidyapeeth, Rahuri, District Ahmednagar.

Besides these, there are three sub-centres in Maharashtra located at Pune, Jalgaon and Padegaon where cotton breeding work is carried.

The main focus of above research centres is to develop productive cultivars of *G. hirsutum* and *G. arboreum* and diploid and tetraploid hybrids. The Rahuri centre is concentrating for development of productive hybrids for irrigated conditions, while other two centres are breeding varieties and hybrids mainly for rainfed conditions. Initially, plant introduction, pure line selection and mass selection methods were used for cotton improvement. Now all breeding techniques, *viz.*, introduction, mass selection, pure line selection, hybridization, mutation and transgenic breeding are used for genetic improvement of cotton.

Progress of Breeding

There has been remarkable progress in cotton breeding in the past. The lint yield increased from 57 kg/ha (in 1975) to 353 kg/ha. Other important achievements of cotton breeding in Maharashtra state are listed below:

1. Development of jassid resistant cultivars [Buri 1007 and DHY 286].
2. Development of short duration cultivars [AKA 7 and PA 255].
3. Development of productive hybrids [both upland and diploid cottons].
4. Improvement in fibre length and spinning capacity.
5. Development of long staple arboreum cultivars [AKA 8401].
6. Identification of GMS line in *G. arboreum* at PDKV Akola centre.
7. Development of CMS based intra-hirsutum hybrids [PKV Hy 3, PKV Hy 4 and PKV Hy 5].
8. Improvement in ginning outturn [AKH 7].
9. Release of productive hybrid of diploid cotton. [Pha 46 and AKDH 7].

Practical Achievements

Remarkable progress has been made in cotton breeding since the inception of AICCIP in 1967. Several varieteis of *G. hirsutum*, *G. arboreum* and hybrids have been released for commercial cultivation in different cotton growing regions (Appendix 1). The varieties which were under cultivation till recently included DHY 286, Rajat, Purnima, AKH 081, LRA 5166, and LRK 516 in *G. hirsutum*; and AKH 4, AKA 5, AKA 8401, AKA 7, Y1, PA 183 and PA 255 in *G. arboreum*. Here, it is essential to make a mention about few varieties. For example, variety AKH 081 is a short duration and semi-dwarf variety of upland cotton. PA 183 is a long staple variety of *G. arboreum* released for Marathwada region. Besides varieties, eleven hybrids have been released for commercial cultivation in Maharashtra state. The hybrids which were under cultivation before introduction of *Bt.* cotton hybrids included NHH 44, PKV Hy2, PKV Hy5, NHH 302, NHB 12, DCH 32 and Pha 46. The last hybrid is an interspecific diploid hybrid released for Marathwada region. Some of the hybrids, *viz.*, H4, H6, H8 and H10 released from Gujarat state also became popular in Maharashtra.

3. GUJARAT STATE

Gujarat is the highest cotton producing state of India. The world's first cotton hybrid known as Hybrid 4 or H 4 was developed for commercial cultivation from Gujarat State by Dr. C. T. Patel. For this outstanding contribution in cotton improvement Dr. Patel was conferred Padamshri. Dr. Patel is rightly called as the father of hybrid cotton and Gujarat is called as the home of hybrid cotton.

Species Composition

In this state, three species of cotton *viz.*, *G. hirsutum*, *G. herbaceum* and *G. arboreum* are cultivated. Earlier hybrid cotton covered maximum area (50 per cent) followed by *G. herbaceum* varieties [25 per cent], *G. hirsutum* [15 per cent] and *G. arboreum* [10 per cent]. Now intra-*hirsutum Bt.* cotton hybrids are pre-dominant covering maximum area. Very little area is under *G. herbaceum* and *G. arboreum* cottons.

Growing Regions

In Gujarat, cotton is cultivated in the entire state. However, the cotton growing area of Gujarat is divided into four major regions, *viz.*, (1) South Gujarat, (2) Middle Gujarat, (3) Mathio Tract, and (4) Wagad Area.

1. South Gujarat

This region includes Surat and part of Bharuch and covers about 4 lakh ha. area. In this region, about 80-90 per cent cotton area is irrigated. The intra-*hirsutum* hybrids are predominant followed by *G. herbaceum*. This region has deep black soil.

2. Middle Gujarat

This includes Baroda, Panch Mahal, part of Bharuch, Sabarkantha and area adjacent to Narmada. In this area, intra-*hirsutum* hybrids and varieties of *G. hirsutum* and *G. herbaceum* are cultivated. In this region, about 50 per cent area is irrigated. The area in Panch Mahal district is mostly rainfed. This region also covers about 4 lakh ha. area.

3. Mathio Tract

This area includes Amreli, Bhavnagar and some talukas of adjacent district. This tract covers 0.5 lakh ha. area which is entirely rainfed. In this region, intra-*hirsutum* hybrids are pre-dominant followed by *G. arboreum* cultivars.

4. Wagad Area

This tract includes North Gujarat and Saurashtra except Mathio Tract including Kutch. The main districts include Mahsana, Banaskantha, Ahmedabad and Surendranagar which cover 25 per cent of Gujarat cotton area. In this tract, cotton is grown on 6-7 lakh ha. area. In this region, 5 per cent well irrigation is available. The *G. herbaceum* is the pre-dominant species followed by intra-*hirsutum* hybrids.

Breeding Centres

In Gujarat, cotton breeding work is carried out at Surat, Talod, Bharuch, Chharodi and Junagarh. Surat is the main cotton research centre and others are sub-centres. The main activity of Surat centre is to develop productive cultivars of *G. hirsutum, G. herbaceum* and *G. arbroeum*, besides intra-*hirsutum* and diploid hybrids. The Bharuch centre is working mainly on improvement of *G. herbaceum*. The world's first cotton hybrid *i.e.* H4 was developed and released from Main Research Station, Surat.

Besides pure line selection, mass selection, inter-varietal hybridization; inter-specific hybridization was used for transfer of desirable characters from one species to another. Some varieties were developed from inter-specific derivatives [Gujarat 67, Deviraj and Devitej].

Progress of Breeding

The important achievements of cotton breeding in Gujarat state are listed below:

1. Development of first cotton hybrid (H4) for commercial cultivation.
2. Development of insect and disease resistant cultivars.
3. Development of short duration cultivars of *G. herbaceum.*
4. Development of productive intra-*hirsutum* hybrids.
5. Development of productive diploid hybrids.
6. Improvement in fibre length and spinning capacity.
7. Improvement in ginning outturn.

Practical Achievements

In Gujarat state, remarkable work has been done in cotton breeding. In this state, several varieties of *G. hirsutum* and *G. herbaceum*, few varieties of *G. arboreum* and several hybrids have been released for commercial cultivation (Appendix 1). The varieties and hybrid which were under cultivation prior to *Bt.* cotton hybrids are given below:

1. *Gossypium hirsutum* Varieties

In this species, important varieties include G. Cot 12, G.Cot 16 and G.Cot 18. Earlier this species coverered about 15 per cent area.

2. G. herbaceum Varieties

In this species, important varieties include G. Cot 13, G. Cot 17, G. Cot 19, G. Cot 21 and G. Cot 23. Earlier this species covered about 25 per cent area.

3. G. arboreum Varieties

In this species, only two varieties *viz.*, G. Cot 15 and G. Cot 19 have been released for cultivation. Very little area is covered 5 per cent) by *G. arboreum* in the state of Gujarat.

4. Hybrids

Four intra-*hirsutum* hybrids and two interspecific diploid hybrids between *G. herbaceum* and *G. arboreum* were released for commercial cultivation. The intra-*hirsutum* hybrids include H 4, H 6, H 8 and H 10. The interspecific diploid hybrids include DH 7 and DH 9. The DH 7 was the first diploid hybrid and DH 9 was the diploid hybrid with long staple. The Hybrid 4 is the world first cotton hybrid which was released from the Main Cotton Research Station, Surat of Gujarat Agricultural University. This hybrid later on spread to several other states such as Karnataka, Andhra Pradesh, Tamil Nadu, Maharashtra and Madhya Pradesh by virtue of its good agronomic performance and wider adaptability.

FUTURE PROSPECTS

In central zone, the future cotton breeding work needs to be directed towards the following thrust areas:

1. About 50 per cent of cotton growing soils are shallow. Hence, there is need to develop *Bt.* varieties and hybrids suitable for shallow soils.

2. About 70 per cent of the cotton crop is grown under rain-fed conditions which suffers from moisture stress (drought) at one or the other stage. There is ample scope to develop drought tolerant *Bt.* varieties and hybrids.

3. The cost of *Bt.* hybrids is very high. Hence, there is need to develop male sterility based *Bt.* hybrids to provide hybrid seed to the farmers at cheaper rate.

4. The cultivation of hybrids is input oriented which may not be afforded by small and marginal farmers. There is need to develop *Bt.* cotton hybrids suitable for low input technology.

5. The fibre strength of varieties and intra-*hirsutum* hybrids is poor. There is need to develop *Bt.* varieties and hybrids with high fibre strength suitable for high speed spinning.

6. Varieties of *G. arboreum* cotton possess high degree of resistance to sucking pests and abiotic stresses. However, they have poor fibre properties. There is ample scope to develop *Bt.* varieties and hybrids of *G. arboreum* cotton with high fibre length and strength.

7. Picking has become a very expensive operation due to very high wages of laborers. There is need to develop *Bt.* cotton varieties and hybrids with synchronous maturity suitable for machine picking.

8. The yield level has reached a plateau. New breeding approaches *viz.*, recurrent selection, bi-parental mating, disruptive mating and diallel selective mating may prove rewarding in achieving further yield improvement.

SUMMARY

In central zone, cotton is grown under rain-fed conditions and sowing is done with the onset of monsoon. In this zone, two species of cotton, *viz.*, *G. hirsutum* and *G. arboreum* and their hybrids are commercially cultivated. The soil moisture stress is the major problem. Several varieties of upland cotton and few varieties of *G. arboreum* and few hybrids were released which were till recently under cultivation. Now 296 *Bt.* cotton hybrids have been approved by the government of India for cultivation in this zone. Now the maximum area is occupied by Intra-hirsutum *Bt.* hybrids and remaining by *G. arboreum* varieties. Some *G. herbaceum* varieties are grown in Gujarat state. Future thrust areas of cotton improvement for Maharashtra State are highlighted.

QUESTIONS

1. Explain main features of cotton cultivation in Central Zone.

2. Discuss briefly main problems of cotton breeding in Central Zone.

3. Describe briefly progress of cotton breeding in Central Zone.

4. Give a brief account of practical achievements of cotton breeding in Central Zone.

5. Discuss briefly future thrusts of cotton breeding in Central Zone.

6. Explain in brief practical achievements of cotton breeding in:

    (i)   Madhya Pradesh   (ii) Maharashtra

    (ii) Gujarat   (iv) Western Maharashtra

7. Give a brief account of cotton hybrids released in Central Zone.

8. Explain the role of Central Institute for Cotton Research in breeding varieties for Central Zone.

9. Describe the contribution of following cotton breeders:

    (i)   C.T. Patel   (ii) M.A. Tayyab

    (iii) N.P. Mehta   (iv) B.H. Katarki

Chapter 8

Breeding in South Zone
[Andhra Pradesh, Telangana,
Karnataka and Tamil Nadu]

INTRODUCTION

South zone includes Andhra Pradesh, Telangana, Karnataka, and Tamil Nadu. Cotton cultivation is done both under irrigated and rainfed conditions. Soils of this zone are both black and red and poor in fertility. In this region intra-*hirsutum* and *hirsutum x barbadense Bt.* hybrids are predominant and some area is under Egyptian cotton and diploid cottons. The area is well known for growing long and extra long staple *G. barbadense* cottons. Cotton is grown in south as sole crop or as intercrop with onion, chilli, cowpea, maize, *etc.* Cotton- rice rotation is also followed in this region. Main points related to cotton cultivation in south zone are as follows:

1. In this zone, cotton crop is grown on black and red soils having undulating topography.

2. In this zone, about 25 per cent of cotton crop is cultivated under irrigated conditions and rest as rain-fed crop.

3. Andhra Pradesh including Telangana covers about 11 per cent of total cotton area in the country and contributes about 17 per cent to the national cotton production.

4. The seed cotton yield ranges from 12-15 q/ha in this zone.

5. In most of the areas, cotton is cultivated as *kharif* crop. In rice fallow areas cotton is also cultivated as rabi crop.

6. The cotton varieties and hybrids grown in south zone belong to medium, superior medium, long and extra-long staple groups.

MAJOR PROBLEMS

The major problems in this zone related to cotton cultivation are listed below:

1. There are inadequate irrigation facilities.
2. There is moisture stress during crop season.
3. In some areas there are shallow soils.
4. There is severe incidence of whitefly in some areas.
5. Some areas have soil salinity and water logging.
6. Infection of Verticillium wilt in upland cotton.
7. Excessive use of pesticides.

1. ANDHRA PRADESH AND TELANGANA

Andhra Pradesh is one of the major cotton growing states of India. It comes under southern cotton growing belt of India. It covers about 11 per cent of total cotton area and contributes 17 per cent to the national cotton production.

Species Composition

There are four cultivated species of cotton *viz., Gossypium hirsutum, Gossypium barbadense* (both tetraploid), *G. arboreum* and *G. herbaceum* (diploids) which are commercially cultivated in India. Andhra Pradesh has privilege of growing all the four cultivated species and tetraploid hybrids of cotton. The maximum area is covered by *Bt.* cotton hybrids (99.5 per cent) and rest by varieties of *G. hirsutum* and *G. arboreum.* Very little area is covered by *G. herbaceum* and *G. barbadense* cottons.

1. Upland Cotton

This is also known as American cotton and is cultivated in all three cotton growing tracts of Andhra Pradesh. This species is successfully cultivated both under rainfed and irrigated conditions.

2. Egyptian Cotton

This is also known as Sea Island cotton or quality cotton. This species is cultivated on small areas only under irrigated parts of Nagarjunasagar Project and Central Tract. It is cultivated only during *kharif* season.

3. Arboreum Cotton

This species is cultivated in the rainfed areas of central and northern tracts. It is not cultivated in the eastern tract, where Upland and Egyptian cottons and tetraploid hybrids are cultivated.

4. Herbaceum Cotton

This species is cultivated in the central tract particularly in the Hingari area as a late *kharif* crop. It is grown in Kurnool and Anantpur distraicts on a very small area.

5. Hybrids

In Andhra Pradesh, tetraploid hybrids are grown in all the three cotton growing tracts. Intra-*hirsutum* hybrids are cultivated both under rainfed and irrigated conditions. However, interspecific hybrids (*hirsutum* x *barbadense*) are cultivated under irrigated conditions in all three cotton growing tracts. Such hybrids cannot be cultivated under rain-fed conditions as they are susceptible to drought conditions.

Growing Regions

In Andhra Pradesh, cotton is cultivated in the entire state. However, the cotton growing area of Andhra Pradesh is divided into three major tracts *viz.,* (1) Eastern tract, (2) Central tract, and (3) Northern tract. These are briefly discussed below :

1. Eastern Tract

This tract includes Nagarjunasagar Project areas and rice fallow areas. This tract has irrigation facilities and, therefore, Egyptian cotton, American cotton and both intra and interspecific tetraploid hybrids are successfully grown. The important cotton growing districts of this tract are Prakashan, Krishna, Guntur and Nellore. In rice fallow areas, cotton is grown in *rabi* season.

2. Central Tract

This tract includes Hingari and Mungari cotton growing areas. This tract is mainly rainfed and very little area is irrigated. In rainfed areas, upland and *desi* cotton varieties are cultivated. In irrigated areas, Egyptian cotton (Suvin) and tetraploid hybrids are cultivated. In this tract, the cotton crop is grown only in *Kharif* season. In the Hingari area, the cotton crop is planted in the late season. This tract includes Anantpur, Cuddapah and Kurnool districts. This tract is also known as Rayalseema area.

3. Northern Tract

This tract includes, Ghat areas of Adilabad district, Gaorani cotton area and Telangana region including Sri Ramsagar Project areas. In Adilabad area, cotton crop is cultivated as rainfed and varieties of both American and *G. arboreum* cotton are cultivated. In the irrigated area of north Telangana, both upland varieties and tetraploid hybrids are cultivated. The important cotton growing districts of Northern tract are : Adilabad, Mehboobnagar, Nalgonda, Khammam, Warrangal, Karimnagar, Nizamabad, Rangareddy and Medak.

Breeding Centres

In Andhra Pradesh, cotton improvement work started in 1904 with the establishment of the Department of Agriculture. This work was further strengthened in 1923 when Indian Central Cotton Committee (ICCC) was constituted. The varietal improvement work got momentum with the inception of All India Coordinated Cotton Improvement Project (AICCIP) in April, 1967.

In Andhra Pradesh, cotton improvement work is carried out at Agricultural Research Stations, Guntur (Lam Farm) and Nandyal which came under Acharya

N.G.Ranga Agricultural University (ANGRAU), Hyderabad. The Lam Farm, Guntur is the main research centre and Nandyal as the sub-centre. The main focus of these two research centres is to develop high yielding varieties of *G. hirsutum* and *G. arboreum* cotton and intra-*hirsutum* hybrids for cultivation in three cotton growing tracts of Andhra Pradesh. The testing work is also carried out at Adilabad and Mudhol centres.

PROGRESS OF BREEDING

In Andhra Pradesh, remarkable achievement has been made in cotton breeding since inception of AICCIP. The major achievements are as follows:

1. Improvement in seed cotton or lint yield.
2. Improvement in fibre quality especially length and spinning capacity.
3. Development of insect and disease resistant varieties and hybrids.
4. Development of early maturing varieties suitable for multiple cropping systems
5. Development of long staple *arboreum* cultivar.

Use of Elite Germplasm

In the germplasm, genetic differences do exist among genotypes for various characters such as earliness, boll size, ginning percentage, fibre length and strength, resistance to insects, diseases, drought, salinity, water logging *etc.* Such elite germplasm lines should be utilized in the breeding programmes for developing productive and resistant varieties and hybrids with good fibre properties. List of some elite germplasm lines of upland and arboreum cottons for various economic characters is presented in Appendix 4.

Practical Achievements

After the inception of AICCIP in 1967, sixteen varieties of *G. hirsutum*, 4 varieties of *G. arboreum* and one variety of *G. herbaceum* have been released/recommended for commercial cultivation in different cotton growing tracts of Andhra Pradesh (Appendix 1). The varieties which were under cultivation prior to introduction of *Bt.* cotton hybrids included Kanchana, LK 861, L 389, L 603, L 604, Sahana, Surabhi, Sumangala, NA 1325, NA 247 and NA 920 in *G. hirsutum*; Srisailam, Mahanandi, Saraswati and Arvinda in *G. arboreum*; Raghvendra in *G. herbaceum*; and Suvin in *Gossypium barbadense*. Here, it is important to make a mention about few varieties. For example, varieties Kanchana and LK 861 are tolerant/resistant to whitefly and can be successfully grown in whitefly prone areas. The variety Saraswati is the first long staple variety of *G. arboreum* released in Andhra Pradesh. The variety Suvin, though released from Central Institute for Cotton Research, Regional Station, Coimbatore, is also cultivated in irrigated areas of Andhra Pradesh. This variety is comparable to GIZA 45 variety of Egypt in its fibre quality.

Besides varieties, hybrids were also released for commercial cultivation in Andhra Pradesh. After the start of AICCIP in 1967, several varieties and hybrids were released for cultivation in Andhra Pradesh (Appendix 1). Most of the hybrids for Andhra Pradesh were released from other states such as Karnataka (Varalaxmi, DCH 32, DHB 105, DHH 11, DDH 2), Tamil Nadu (Savita, Surya, Sruthi), Gujarat (H6) and Madhya Pradesh (JKHY 1). Three hybrids *viz.,* NHB 80, LAHH 1 and LAHH 4 were released from Cotton Research Centre, Guntur of Andhra Pradesh. Now 294 *Bt.* cotton hybrids have been released for cultivation in all the three cotton growing States of south zone.

2. KARNATAKA STATE

Karnataka is one of the major cotton growing states of India which comes under the southern zone. It covers about 6.5 per cent of total cotton area in the country and contributes 5.8 per cent to the national production.

Species Composition

In Karnataka, all the four cultivated species of cotton *viz., G. hirsutum, G. arboreum, G. herbaceum, Gossypium barbadense* and hybrids are cultivated on commercial scale. The maximum area is covered by *Bt.* hybrids followed by diploid cottons (4 per cent) and *G. hirsutum* varieties (3 per cent). Very little area is covered by *Gossypium barbadense* cotton.

Growing Regions

In Karnataka, cotton is grown in the entire state. However, the cotton growing area of Karnataka State is divided into three major regions, *viz.,* (1) irrigated tract, (2) rainfed tract, and (3) western dry zone.

Research Centres

In Karnataka, cotton improvement work is carried out at Dharwad, Arbhavi and Siruguppa. Dharwad is the main research centre and rest are sub-centres. The main focus is to develop productive cultivars of *G. hirsutum* and *G. herbaceum* and tetraploid and diploid hybrids. The world's first interspecific hybrid between *G. hirsutum* and *Gossypium barbadense* was released from Dharwad Research Centre which now comes under University of Agricultural Sciences, Dharwad.

Progress of Breeding

Remarkable improvement has been achieved in cotton breeding in the past in Karnataka (Appendix 1). The major achievements are as follows:

1. Development of first interspecific hybrid *i.e.* Varalaxmi between *G. hirsutum* and *Gossypium barbadense.*
2. Improvement in seed cotton or lint yield.
3. Improvement in fibre quality especially length and spinning capacity.
4. Development of insect and disease resistant varieties and hybrids.
5. Development of early maturing varieties suitable for multiple cropping systems.
6. Development of productive intra and interspecific tetraploid hybrids.
7. Development of productive diploid hybrid for rainfed cultivation.

Practical Achievements

In Karnataka, remarkable work has been done in cotton breeding. In this state, 10 varieties of *G. hirsutum* and seven of *G. herbaceum* have been released so far. The cultivated varieties include Sharda, Abadhita and Sahana in *G. hirsutum*; DB 3-12 and Raichur-51 in *G. herbaceum*. Besides these varieties, four hybrids in tetraploid cotton *viz.,* Varalaxmi, DCH-32, DHB-105 and DHH-11 and one of diploid cotton *viz.,* DDH-2 have been released for commercial cultivation in different area (Appendix 1). The presently cultivated hybrids include DCH-32, DHB-105, DHH-11 and DDH-2.

3. TAMIL NADU STATE

Tamil Nadu is one of the major cotton growing states of India which comes under the southern zone. It covers 1.27 per cent of total cotton area in the country and contributes 1.66 per cent to the National production.

Species Composition

In Tamil Nadu, three species of cotton *viz., G. hirsutum, G. arboreum* and *Gossypium barbadense* and hybrids are cultivated on commercial scale. The maximum area is covered by tetraploid *Bt.* cotton hybrids[83 per cent] and remaining by *G. hirsutum* non-Bt hybrids [12 per cent] and *arboreum* varieties (5 per cent). Very little area is covered by *G. barbadense* cotton.

Growing Regions

In Tamil Nadu, cotton crop is grown in the entire state. However, the cotton growing area of Tamil Nadu is divided into four major regions, *viz.,* (1) Winter Cambodia Region, (2) Summer Cambodia Region, (3) Karungani Tract, and (4) Rice Fallows.

Reserch Centres

In Tamil Nadu, cotton breeding work is carried out at the following research centres:

1. Cotton Section, Tamil Nadu Agricultural University, Coimbatore.
2. Agricultural Research Station of TNAU, Kovilpatti.
3. Agricultural Research Station of TNAU, Srivilliputtur.
4. Central Institute for Cotton Research, Regional Station, Coimbatore.

The Coimbatore is the main centre and others are sub-centres. The main activity of Coimbatore centre is to develop superior cultivars in *G. hirsutum* and *Gossypium barbadense* and also to develop productive intra-*hirsutum* and interspecific hybrids between these two species. The main objective of Kovilpatti centre is to develop productive cultivars of *G. arboreum* for rainfed areas of Tamil Nadu.

Progress of Breeding

Remarkable achievement has been in cotton breeding in the past in Tamil Nadu (Appendix 1). The major achievements are:

1. Improvement in seed cotton or lint yield.
2. Improvement in fibre quality especially length and spinning capacity.
3. Development of insect and disease resistant varieties and hybrids.
4. Development of early maturing varieties suitable for multiple cropping systems.
5. Development of productive intra and interspecific tetraploid hybrids.
6. Development of *Gossypium barbadense* variety for commercial cultivation.

Practical Achievements

In Tamil Nadu, remarkable work has been done in cotton breeding. In this state, 25 varieties of *G. hirsutum,* three of *Gossypium barbadense* and seven of *G. arboreum* have been released (Appendix 1). The main cultivated varieties include MCU 5 VT, MCU 7, LRA 5166, SVPR 2, Surabhi and Sumangala in *G. hirsutum;* K 10 and K 11 in *G. arboreum* and Suvin in *Gossypium barbadense.* Besides these varieties, several hybrids have been released for commercial cultivation. The main cultivated hybrids include Savita, TCHB 213, Surya and Sruthi.

FUTURE THRUSTS

In this zone, remarkable improvement in productivity has been achieved during the past few decades. To achieve further improvement in yield, the future breeding efforts need to be directed towards the following thrust areas.

1. In this zone, *Bt.* hybrids cover vast cotton area. Hence, efforts should be made to develop productive male sterility based *Bt.* cotton hybrids to provide hybrid seed at cheaper rate to the farmers.
2. In Andhra Pradesh, the cotton crop suffers from severe incidence of whitefly causing heavy yield losses. Varieties with smooth leaf and stem are resistant to whitefly. At present, two varieties of upland cotton *viz.,* LK 861 and Kanchana having resistance to whitefly are available. However, yield potential of these varieties is poorer than susceptible varieties. Hence, there is need to develop productive *Bt.* cultivars of upland cotton with resistance to whitefly. The resistant lines for whitefly can be utilized from the cotton germplasm available at CICR, Nagpur.
3. Varieties of *G. arboreum* are resistant or tolerant to sucking pests including whitefly. Hence, efforts should be made to develop productive *Bt.* cultivars of *G. arboreum* cotton with good fibre properties for whitefly prone areas.
4. In this zone, about 70 per cent of cotton crop is grown as rain-fed which often suffers from moisture stress at one or the other stage. Hence, there is need to develop drought tolerant *Bt.* cultivars and hybrids of upland cotton.
5. In some areas, there is problem of soil salinity and water logging. For such areas, there is need to develop *Bt.* varieties resistant to soil salinity and water logging conditions.

6. Under rain-fed areas, about 50 per cent of the soils are shallow. There is need to develop varieties suitable for cultivation in shallow soils.

7. Fibre strength is an important character for high speed spinning *i.e.* jet spinning and rotobar spinning. In the international market, fibre strength is used as an important criterion in fixing the price of cotton. Hence, efforts should be made to develop *Bt.* cotton varieties and hybrids with high fibre strength to fetch higher price in international market.

8. Now, labor is becoming costlier day by day. Thus, manual picking of cotton has become very expensive. There is need to develop *Bt.* cotton varieties and hybrids with synchronous boll opening suitable for machine picking.

9. In this zone, only one variety of Egyptian cotton *i.e.* Suvin is grown for last more than four decades. There is need to develop variety of Egyptian cotton better than Suvin especially in yield and ginning percentage.

10. Majority of farmers in rain-fed areas are poor and they cannot use high inputs especially fertilizers and pesticides in cotton crop. Hence, there is need to develop varieties and hybrids suitable for low input technology.

11. Conventional breeding procedures *viz.,* pure line selection, mass selection and pedigree breeding have become ineffective in achieving further improvement in yield. Hence, various population improvement approaches such as recurrent selection, disruptive mating, bi-parental mating and diallel selective mating may be rewarding in creating further variability in the population to make effective selection.

SUMMARY

In this zone, cotton is mainly grown under rain-fed [70 per cent] conditions and sowing is done with the onset of monsoon. Several varieties of upland cotton and few varieties of *G. arboreum* and *G. herbaceum* and few hybrids were released which were till recently under cultivation. Now 294 *Bt.* cotton hybrids have been approved by the government of India for cultivation in south zone. Now the maximum area is occupied by Intra-*hirsutum Bt.* hybrids and remaining by *G. arboreum* varieties. Future thrust areas of cotton improvement for south zone have been highlighted.

QUESTIONS

1. **Explain main features of cotton cultivation in South Zone.**

2. **Discuss briefly main problems of cotton breeding in South Zone.**

3. **Describe briefly progress of cotton breeding in South Zone.**

4. **Give a brief account of practical achievements of cotton breeding in South Zone.**

5.   Discuss briefly future thrusts of cotton breeding in South Zone.

6.   Explain in brief practical achievements of cotton breeding in:

(i)   Andhra Pradesh   (ii) Karnataka

(iv) Tamil Nadu   (iv) Telangana

7.   Give a brief account of cotton hybrids released in South Zone.

8.   Explain the role of Central Institute for Cotton Research in breeding varieties for South Zone.

9.   Describe the contribution of following cotton breeders:

(i)   V. Santhanam   (ii) B.H. Katarki

(iii) R. Krishna Moorthy   (iv) N.G.P. Rao

Chapter 9

Transgenic Breeding

INTRODUCTION

A cotton genotype developed by the technique of plant biotechnology is referred to as transgenic cotton. In other words, genetically engineered cotton is known as transgenic cotton. A cotton genotype that contains gene from the soil bacterium *Bacillus thuringiensis* is called *Bt.* cotton. It may be a pure breeding variety or a hybrid. Genetic improvement of crop plants through the application of agricultural biotechnology is referred to as transgenic breeding. When it is applied to cotton improvement, it is called transgenic cotton breeding.

TRANSGENIC PLANTS

The genetically engineered plants are called transgenic plants. Main features of transgenic plants are listed as under.

(i) Transgenic plants contain transgene or foreign gene of special significance.

(ii) Transgenic plants are developed through plant biotechnology.

(iii) In the development of transgenic plants, the sexual process is not involved.

(iv) The transgenic plants are recovered at a very low frequency.

(v) Transgenic plants are developed to solve those problems which cannot be solved by conventional plant breeding techniques.

ADVANTAGES OF TRANSGENIC TECHNOLOGY

Transgenic technology has several advantages such as (i) rapid method, (ii) free gene transfer, (iii) single gene transfer, (iv) direct gene transfer and (v) solution to difficult problems. These are briefly discussed as follows:

(i) Rapid Method

It is an effective and rapid method of crop improvement. It takes 3-4 years for developing new cultivars/hybrids against 10-15 years taken by conventional methods.

(ii) Free Gene Transfer

It permits gene transfer across the species, genera, family even from unrelated organisms.

(iii) Single Gene Transfer

Gene technology permits transfer of one or two genes from donor parent to the recipient parent. In conventional hybridization method, hundreds of genes are transferred to the recipient parent. Many of transferred genes are undesirable. Elimination of such genes requires repeated backcrossing to the recipient parent which takes 4-5 years.

(iv) Direct Gene Transfer

Gene technology permits direct gene transfer into the recipient parent bypassing sexual process. In other words, there is no need of union of male and female gametes in gene technology. The gene of interest can be directly inserted into the cell of recipient parent.

(v) Solution to Difficult Problems

Gene technology provides solution to those problems that cannot be solved by conventional methods of breeding. The best example is resistance to bollworms in cotton.

APPLICATIONS OF TRANSGENIC TECHNOLOGY

Transgenic technology has several practical applications. Important applications of transgenic technology include improvement in (i) biotic resistance, (ii) abiotic resistance, (iii) herbicide resistance, (iv) quality of food products, (v) development of novel traits such as golden rice and male sterility, (vi) industrial products and (vii) bioremediation. All these aspects are briefly discussed as follows:

(i) Resistance to Biotic Stresses

Biotic stress refers to adverse effects on crop growth and yield by biotic factors such as insects, diseases and parasitic weeds. In crop plants, heavy yield losses are caused every year due to insect and disease attack. Moreover, insecticides and pesticides which are used to control insects and diseases are expensive and have adverse effects on other beneficial organisms (parasites and predators). Gene technology has played key role in developing insect resistant cultivars in several crops such as bollworm resistant cultivars in cotton and stem borer resistant cultivars in maize. Moreover, the technology is eco-friendly.

(ii) Resistance to Abiotic Stresses

Abiotic stress refers to adverse effects on crop growth and yield by abiotic factors such as drought, soil salinity, soil acidity, cold, frost, *etc*. The cold resistant genotypes in tobacco and freezing resistant genotypes have been developed through gene technology. Efforts are being made to develop drought and salinity resistant cultivars in many crops.

(iii) Herbicide Resistance

In crop plants, weeds cause heavy yield losses and also adversely affect quality of the produce. The genetic resistance is the cheapest and the best way of solving this problem. Gene technology has been used to develop herbicide resistant cultivars in cotton, maize, wheat, tobacco, potato, tomato, rapeseed, soybean, flax, *etc*. In these crops, cultivars resistant to glyphosate, gluphosinate and some other herbicides have been developed.

(iv) Improvement in Quality

The quality is adjudged in three ways, *viz.*, nutritional quality, market quality [keeping quality] and industrial quality. Gene technology has helped in improving these qualities in different crops. For example, the ripening and softening in tomato has been delayed. It is desirable for safe transport and storage. This has been achieved by manipulating the genes that encode the enzyme responsible for ripening [ethylene forming enzymes and softening [polygalactonase]. Other quality improvement include non-browning potato, starch composition of wheat flour, carotene content in rice and improved oil content in oilseed crops.

(v) Development of Novel Traits

Plant biotechnology plays important role in development of novel traits. Examples are golden rice and transgenic male sterility in rapeseed.

(vi) Industrial Products

Gene technology has great potential for the production of biodegradable plastics, obtaining therapeutic proteins, pharmaceuticals and edible vaccines from transgenic plants. It may also help in producing biodiesel or petroleum products.

(vii) Bioremediation

It is possible to develop plants which can be used for bio-remediation of sick soils.

NEED FOR *Bt*. COTTON

In cotton bollworms cause significant yield losses. Three types of bollworms, *viz.*, American bollworm (*Helicoverpa armigera*), pink bollworm (*Pectinophora gossypiella*) and spotted bollworms (*Earias vitella*) attack cotton crop. We do not have any source of resistance to the bollworms in the germplasm of cotton including wild species and wild relatives the world over. Moreover, about 10 per cent of insecticides on global basis and 45 per cent in India are used for control of insects in cotton crop alone.

ADVERSE EFFECTS OF INSECTICIDES

There are several adverse effects of Insecticides. The main adverse effects include killing of (i) parasites and predators, (ii) beneficial insects, (iii) beneficial micro-organisms, (iv) environmental pollution, (v) increase in the cost of cultivation, (vi) development of resistance in the insects, and (vii) adverse effects on animal and human health. These are briefly discussed as follows:

(i) Killing of Predators and Parasites

There are several natural predators and parasites of bollworms, which feed on eggs and larvae of bollworms such as *Crysoperla* and lady bird beetle to cite a few. The insecticides kill all these insects along with bollworms and thus disturb the ecological balance. The *Bt.* gene, on the other hand, is specific to bollworms and does not have adverse effects on parasites and predators.

(ii) Killing of Beneficial Insects

Besides parasites and predators, there are some other beneficial insects such as honey bee, silkworm, lacworm, *etc*. The application of insecticides kills all these beneficial insects. The *Bt.* gene does not have any adverse effect on such insects.

(iii) Killing of Micro-organisms

There are several microorganisms such as earthworm, blue green algae, and nitrogen fixing bacteria which are beneficial for agriculture. The use of insecticides kills all these beneficial micro-organisms. The *Bt.* gene does not have any adverse effect on such micro-organisms.

(iv) Environmental Pollution

The use of insecticides leads to environmental pollution such as soil pollution and water pollution. The polluted water goes to rivers, ponds and lakes along with rain water. When animals drink such polluted water it has adverse effect on their health. The high level of insecticides in the water sometimes leads to death of the animals. The *Bt.* gene does not have any such adverse effect.

(v) Increase in Cost of Cultivation

The indiscriminate use of insecticides increases the cost of cultivation and make the crop cultivation unprofitable. In other words, much use of insecticides reduces the net profit of the growers.

(vi) Development of Resistance

The too much use of insecticides sometimes leads to development of resistance in the insects against such insecticides. This may lead to heavy yield losses.

(vii) Adverse Effect on Human Health

The insecticides have adverse effects on human health in two main ways. Firstly, they have adverse effects on the people who are engaged in the spraying operations. Secondly, the insecticides go to the food chain in three ways, *viz.*, (i) through drinking water, (especially the well water), (ii) through use of vegetables and fruits sprayed with insecticides, and (iii) through milk and meat of animals which feed on fodders crops sprayed with insecticides. *Bt.* gene does not have any such adverse effect. There is no known source of resistance to bollworms in the cotton genetic resources. The *Bt.* cotton is the only way to control bollworms.

TABLE 9.1: Landmarks in the History of Transgenic Breeding

Year	Name of Crop	Development/Identification of	Country
1983	Tobacco	First transgenic plant in tobacco.	USA
1987	Cotton	First transgenic plant in cotton.	USA by Monsanto, Delta and Pine Companies.
1994	Tomato	FLVR-SAVR tomato.	USA
1994	Potato	First transgenic potato.	USA
1995	Soybean	Herbicide resistant soybean.	USA
1996	Rapeseed	Herbicide resistant Rapeseed.	USA
1996	Corn	Stem borer resistant corn.	USA
1996	Squash	Mosaic resistant Squash.	USA
1997	Technology	Identification of terminator gene.	USA by Monsanto seed company
1998	Technology	Identification of traitor gene.	USA by Monsanto seed company
1999	Papaya	Mosaic resistant Papaya.	USA
2000	Squash	Mosaic resistant Squash.	China
2000	Rice	Golden Rice.	Switzerland
2004	Alfalfa	High protein alfalfa.	USA
2004	Linseed	Herbicide resistant Linseed.	USA
2005	Rice	Golden Rice 2.	USA
2009	Sugar beet	Herbicide resistant Sugar beet.	Canada
2012	Banana	Virus resistant Banana.	Australia and Africa
2013	Corn	Drought resistant Corn.	USA
2013	Sugarcane	Drought resistant Sugarcane.	Indonesia
2015	Apple	Delayed browning apple.	USA and Canada
2015	Beans	Virus resistant bean.	Brazil
2015	Potato	Late blight resistant Potato.	USA

Bt. GENES IDENTIFIED

Some *Bt.* Genes resistant to bollworms were identified in Japan, USA and India between 1990 and 2002. A list of important genes identified by different Seed Companies/public sector organizations so far is presented here under.

TABLE 9.2: List of some *Bt.* Genes Identified by different Organizations

Year	Bt Gene Identified	Identified by
1990	Cry I Ab	Sumitomo company of Japan.
1992	VIP 3 A	Syngenta India Limited.
1993	Cry I Ac	Monsanto Company of U.S.A.
1995	Cry I I Ab	Monsanto Company of U.S.A.
2002	Cry I Aa 3	National Botanical Research Institute, Lucknow.
2002	Cry I A 5	International Centre for Genetic Engineering, New Delhi.
2002	Cry I F	National Research Centre for Plant Biotechnology, IARI, New Delhi.

BT. COTTON IN INDIA

The Chronology of the development of *Bt.* Cotton in India is briefly presented in Table 9.3.

TABLE 9.3: Major Activity Related to *Bt.* Cotton in India

Year	Major Activity Related to Bt. Cotton
March 10, 1995	Department of Biotechnology (DBT), Government of India permitted Mahyco Seed Company to import 100 gm. seed of transgenic cotton variety Cocker-312 cultivated in the United States. This variety contained the Cry 1 Ac gene from the bacterium *Bacillus thuringiensis*.
April 1998	Department of Biotechnology (DBT) permitted Monsanto Seed Company to conduct small scale trials of *Bt.* cotton using 100 g seed per trial.
January 8, 1999	The Review Committee on Genetically Modified organisms [RCGM] reviewed the results of trials conducted at 40 locations and expressed satisfaction.
2000-2002	Trials of *Bt.* cotton were conducted at different All India Coordinated Cotton Improvement [AICCIP] centres of the Indian Council of Agricultural Research in Central and South Zones.
February 20, 2002	The Indian Council of Agricultural Research (ICAR) submitted a positive report to the Ministry of Environment on the field trials of *Bt.* cotton. Based on this report the Genetic Engineering and Approval Committee (GEAC) of the environment ministry approved commercial use of *Bt.* cotton.
March 25, 2002	Government of India gave approval for commercial cultivation of three *Bt.* Cotton hybrids of Mahyco. Seed Company approved by GEAC.
2004	Genetic Engineering and Approval Committee [GEAC] gave approval for commercial cultivation of one hybrid i. e. RCH 2 Bt.
2005	Genetic Engineering and Approval Committee [GEAC] gave approval for commercial cultivation of Sixteen *Bt.* cotton hybrids.
2006	Events *cry 1 Ac* (Event 1); *cry 1 Ac* + *cry 2 Ab* (MON 15985 Event); *cry 1 Ab+cry 1 Ac* were introduced.
2006	Genetic Engineering and Approval Committee [GEAC] gave approval for commercial cultivation of additional 42 *Bt.* cotton hybrids.
2007	Event *cry 1 Ab* and *cry 1 Ac* (GFM *Cry 1A*) were introduced.
2007	Genetic Engineering and Approval Committee [GEAC] gave approval for commercial cultivation of additional 68 *Bt.* cotton hybrids. In total 131 *Bt.* cotton hybrids were approved between 2002-07 for commercial cultivation in India.
2008	Government of India **has** approved commercial cultivation of Bikaneri Narma *Bt.* Cotton in India.
2009	Government of India **has** approved commercial cultivation of *Bt.* NHH 44 hybrid in central India.

TYPES OF *Bt.* COTTON

The transgenic cotton can be divided into different categories on the basis of (i) resistance, (ii) genes involved, (iii) method of development, (iv) material used and (v) species used (Table 9.4).

Base on resistance *Bt.* Cotton is of two types, *viz.*, (1) bollgard cotton, and (2) roundup ready cotton. These are briefly discussed as under:

TABLE 9.4: Types of *Bt.* Transgenic Cotton

Sl.No.	Basis of Classification	Types of Bt. Cotton
(i)	**Resistance**	
	Resistant to bollworm	Bollgard cotton
	Resistance to herbicide	Roundup Ready cotton
(ii)	**Genes involved**	
	Single gene	Bollgard I
	Two genes	Bollgard II
(iii)	**Method of development**	
	Biotechnology	Primary *Bt.* Cotton
	Back cross method	Secondary *Bt.* cotton
(iv)	**Material used**	
	Simple variety	*Bt.* Variety
	Hybrid	*Bt.* Hybrid
(v)	**Species used**	
	Hirsutum	Intra-*hirsutum* hybrids
	Hirsutum x barbadense	Inter-specific hybrids

1. Bollgard Cotton

The transgenic *Bt.* Cotton which is resistant to bollworms is referred to as bollgard cotton. The main features of bollgard cotton are as follows:

  (a) The Bollgard cotton confers resistance to bollworms.

  (b) It is of two types *viz.*, Bollgard I and Bollgard II.

(i) Bollgard I

The *Bt.* Cotton which contains one *Bt.* gene only in its genome is referred to as bollgard I. It generally contains Cry I A (c) gene. The main features of bollgard I cotton are given as under.

  (a) It contains one *Bt.* gene in its genome. In other words, it is a monogenic resistance.

  (b) It generally contains Cry I A (c) gene.

  (c) It is lesser active than Bollgard II cotton.

  (d) It has more chances of breaking resistance.

(ii) Bollgard II

The *Bt.* Cotton which contains two *Bt.* genes in its genome is referred to as bollgard II. In other words, bollgard II contains a combination of two genes in its genome. It generally contains Cry I A © + Cry2 A(b) genes. The main features of bollgard II cotton are given as under.

(a) It contains two *Bt.* genes in its genome. In other words, it is a oligogenic resistance.

(b) The Bollgard II is the second generation transgenic *Bt.* cotton.

(c) It generally contains Cry I A © + Cry2 A(b) genes.

(d) It is 4-5 times more active than Bollgard I cotton.

(e) It has lesser chances of breaking resistance.

(f) Bollgard II is expected to control the entire bollworm complex in cotton.

The bollgard cotton is cultivated in U.S.A., China, Australia, Argentina, Mexico, South Africa, India and some other countries.

2. Roundup Ready Cotton

The transgenic *Bt.* cotton which is resistant to herbicides is referred to as roundup ready cotton. It is resistant to herbicides which are commonly used for control of weeds in cotton crop. In cotton, transgenic plants resistant to various herbicides such as glyphosate, bromoxynil, 2-4D have been developed. The main features of roundup ready cotton are given as follows:

(i) It is a monogenic resistance to herbicides.

(ii) The chances of breaking herbicide resistance in this cotton are more.

(iii) Herbicide resistant cotton is grown in U.S.A. and some other countries.

Based on the method of development, the *Bt.* cotton is again of two types, *viz.*, (1) primary *Bt.* cotton and (2) secondary *Bt.* cotton. These are defined as follows:

1. Primary Bt. Cotton

The *Bt.* cotton which is developed by the technique of genetic engineering bypassing sexual fusion is referred to as primary *Bt.* cotton. Main features of primary *Bt.* cotton are presented as follows:

It is developed by the technique of genetic engineering.

Its development bypasses sexual fusion.

Examples of primary *Bt.* cotton include bollgard cotton and roundup ready cotton.

(i) Its development requires well equipped laboratory and well trained manpower.

(ii) It is an expensive method of developing transgenic cotton.

2. Secondary Bt. Cotton

The *Bt.* cotton which is developed by the conventional breeding methods using one of the parents from primary *Bt.* cotton is referred to as secondary *Bt.* cotton. Main features of secondary *Bt.* cotton are presented as follows:

(i) It is developed by conventional breeding methods. Usually back cross breeding method is used for transfer of *Bt.* gene from primary *Bt.* cotton to new genotypes.

(ii) Its development involves sexual fusion.

(iii Its development requires field and laboratory and lesser technical skill than development of primary *Bt.* cotton.

(iv) It is cheaper method of developing transgenic cotton.

Based on material used, the *Bt.* cotton is of two types, *viz.*, (1) simple variety, and (2) Hybrids. These are explained as follows:

1. Simple Variety

When the *Bt.* gene is incorporated into the background of a straight variety it referred to as simple *Bt.* variety or *Bt.* variety. The straight *Bt.* cotton varieties have been developed for commercial cultivation in countries like USA, Mexico, Australia, Argentina, *etc.* The main advantage of simple varieties is that the seed can be used for many years by the farmers and there is no need of purchasing fresh seed by the farmers every year.

2. *Bt.* Hybrids

When the *Bt.* gene is incorporated into the background of a hybrid variety it referred to as *Bt.* hybrid. The hybrids have been developed for commercial cultivation in countries like India, China, Pakistan, *etc.* The main demerit of *Bt.* hybrids is that the farmers have to purchase fresh seed every year. The *Bt.* cotton hybrids are of two types, *viz.*, (i) intra-specific hybrids, and (ii) inter-specific hybrids. These are explained as follows:

(a) Intra-specific Bt. Hybrids

This group includes hybrids between two different genotypes of the same species. In cotton, intra-specific hybrids have been developed in *G. hirsutum* species. The majority of hybrids released in India belong to this group. The main features of intra-hirsutum *Bt.* hybrids are as follows:

(i) Such hybrids have wider adaptability and can be grown under both rain-fed and irrigated conditions.

(ii) Such hybrids have more tolerance to biotic and abiotic conditions.

(iii) Such hybrids have broad genetic base.

(iv) Such hybrids have lesser problems of neps and motes.

(b) Inter-specific Bt. Hybrids

This group includes hybrids between two different species of the same genus. In cotton, inter-specific hybrids have been developed between *G. hirsutum* and *Gossypium barbadense*. Limited number of hybrids have been released in India in this group. The main features of inter specific *Bt.* hybrids are as follows:

(i) Such hybrids have narrow adaptability and can be successfully grown under irrigated conditions only.

(ii) Such hybrids have lesser tolerance to biotic and abiotic conditions.

(iii) Such hybrids have narrow genetic base.

(iv) Such hybrids have more problems of neps and motes.

TRANSGENIC EVENTS

In genetic engineering, an event is the insertion of a particular piece of foreign DNA into the chromosome of the recipient crop plant. The main points related to an event are listed below:

(i) The Insertion of foreign gene occurs in random manner. In other words, gene integration takes place in random location.

(ii) Each event is unique in its feature.

(iii) The event can affect how a gene is expressed in the organism.

(iv) Once an event occurs, the transgene can be passed to the next generation as a normally inherited gene.

All *Bt.* cotton genotypes till 2009 [hybrids + Variety] have been developed using six different events [Table 9.5].

TABLE 9.5: Details of Six Commercialized *Bt.* Cotton Events in India

Sl.No.	Event	Developed by	Date of Approval
1	MON-531	Mahyco/Monsanto	2002
2	MON-15985	Mahyco/Monsanto	2006
3	Event-1	IIT Kharagpur	2006
4	GFM Event	Chinese Academy of Sciences	2006
5	BNLA-601	CICR (Nagpur) and UAS Dharwad	2008
6	MLS-9124	Metahelix Life Science	2009

1. MON-531 Event

This event was developed by Mahyco/Monsanto. This is known as bollgard I [BG I] which has been used for development of 180 *Bt.* hybrids till 2009. This event used Cry 1 Ac gene. Thus, the *Bt.* cotton containing Cry 1 Ac gene was termed as Bollgard I. The main features of bollgard I cotton are given as under:

(i) It contains one *Bt.* gene in its genome. In other words, it is a monogenic resistance.

(ii) It contains Cry 1 Ac gene.

(iii) It is lesser active than Bollgard II cotton.

(iv) It has more chances of breaking resistance.

(v) it was approved for commercial use in 2002.

2. MON-15985 Event

This event was also developed by Mahyco/Monsanto. It utilized two genes Cry1Ac and Cry2Ab and the hybrids were termed as bollgard II. This event has been used for development of 248 *Bt.* cotton hybrids till 2009. Thus, the *Bt.* cotton which contains two *Bt.* genes [cry1Ac and cry2Ab] in its genome is referred to as bollgard II. In other words, bollgard II contains a combination of two genes in its genome. The main features of bollgard II cotton are given as under:

(i) It contains two *Bt.* genes in its genome. In other words, it is a oligogenic resistance.

(ii) The Bollgard II is the next generation transgenic *Bt.* cotton.

(iii) It contains Cry 1 Ac + Cry2 Ab genes.

(iv) It is 4-5 times more active than Bollgard I cotton.

(v) It has lesser chances of breaking resistance.

(vi) Bollgard II is expected to control the entire bollworm complex in cotton.

(v) It was approved for commercial use in 2006.

3. Kharagpur Event

This event is also known as Event-1. It was developed by Indian Institute of Technology, Kharagpur and used by J.K. Agri-Genetics Pvt. Ltd. The main features of this event are given below:

(i) This event was developed by IIT, Kharagpur.

(ii) It contains Cry1 Ac truncated gene.

(iii) It was approved for commercial use in 2006.

(iv) It was first used by J.K. Agri-Genetics Pvt. Ltd.

The use of this event resulted in the development of 4 hybrids in 2006, 15 hybrids in 2008 and 27 hybrids in 2009.

4. GFM Event

This event was developed by Chinese Academy of Sciences and used by Nath Seeds Ltd. The main features of this event are given below.

(i) It contains fused genes Cry1Ab and Cry1Ac.

(ii) It was first approved for sale in 2006.

(iii) It was first used by Nath Seeds Ltd.

(iv) Initially 3 hybrids were developed in 2006, the number increased to 63 in 2009.

5. Dharwad Event

This event [BNLA-601] was developed jointly by Central Institute for Cotton Research, Nagpur and University of Agricultural Sciences, Dharwad. The main features of this event are given below.

(i) This event contains truncated Cry1Ac gene.

(ii) It was approved for commercial use in 2008.

(iii) This event has been used for development of one *Bt.* hybrid NHH 44 and one *Bt.* cotton variety BN *Bt.*

6. MLS-9124 Event

This event was developed by Indian seed company Metahelix Life Sciences. The main features of this event are given below.

(i) It contains a synthetic gene Cry1C.

(ii It was approved for commercial use in 2009.

(iii) This event has been used for development of two hybrids [MH 5125 and MH-5174] till 2009.

PRACTICAL ACHIEVEMENTS

In cotton, two significant achievements have been made so far through application of biotechnology. These are: (1) bollworm resistance, and (2) herbicide resistance.

1. Bollworm Resistance

In cotton, transgenic plants with resistance to bollworms have been developed. Bollworms cause heavy yield losses in cotton and genetic source of resistance is not available both in cultivated and wild germplasm. The resistant gene has been transferred from soil borne bacterium *Bacillus thuringiensis,* by Monsanto Company in U.S.A. Studies conducted in USA and elsewhere on transgenic cotton indicated that *Bt.* gene was effective in controlling *Helicoverpa zea.* Transgenic cottons expressed control of *Helicoverpa zea* equal to pyrethroid protected non-transformed varieties. There was 93-99 per cent lesser infestation of pink bollworm in transgenic lines than control Coker 312. Transgenic cottons provided excellent control of *Helicoverpa zea* and tobacco budworm (*Heiothis virescens*) in Mississippi delta. Transgenic cottons with *Bt.* endotoxin protein reduced input of insecticides and created eco-friendly environment without reducing crop production. Several bollworm resistant cultivars have been developed and are under cultivation in countries like USA, China, Australia, Africa and Mexico. In India, 521 bollworm resistant *Bt.* cotton hybrids and one variety have been released.

2. Herbicide Resistance

In Cotton, weeds pose a serious problem. The manual cleaning of weeds is a very expensive method. The cheapest and the most practical method of weed control, in the mechanized agriculture, is the use of herbicides. However, herbicides have adverse effects on cotton plants. Biotechnology has helped in developing transgenic plants in cotton, which are resistant to herbicides. In cotton, transgenic plants resistant to various herbicides such as glyphosate, bromoxynil, 2-4D have been developed. In India, *Bt.* cotton hybrids resistant to herbicides have not been released so far.

TABLE 9.6: Details of *Bt*. Cotton Hybrids Released in India till 2009

Sl.No.	Hybrids/Zone/Events	Number of Hybrids
1	Bollgard I	273
2	Bollgard II	248
3	Variety	01
	Total	**522**
4	HH Hybrids	507
5	H x B Hybrids	14
6	Upland Variety	01
	Total	**522**
7	North Zone	164
8	Central Zone	296
9	South Zone	294
	Total	**754***
10	Mon-531 [BG I]	180
11	Mon 15985 [BG II]	248
12	Event 1	27
13	GFM Cry1A	63
14	Dharwad Event	2
15	Event 9124	2
	Total	**522**

Note: *Some hybrids are common across the zone.

LIMITATIONS OF *Bt*. COTTON

The *Bt*. cotton has several advantages such as very effective control of bollworms, reduction in pesticide, reduction in cost of cultivation, eco-friendly, *etc*. As a result, it has replaced non *Bt*. cotton on vast area. Despite several advantages, there are some limitations of *Bt*. cotton. Major limitations of *Bt*. cotton include effectiveness up to 120 days, ineffective against sucking pests, adverse effect on insecticide industries and adverse effect on employment. It also promotes malpractices such as mixing of non-*Bt*. seed with *Bt*. and sale of F_2/F_3 seeds. These aspects are discussed below:

1. Effectiveness up to 120 Days

The toxin producing ability of the *Bt*. gene integrated in the cotton genome ranges from 90–120 days depending upon the maturity duration of the *Bt*. hybrid. In early maturing hybrids, it is effective up to 90 days, in medium maturing hybrids up to 100 days and in late maturing hybrids up to 120 days. After that, the toxin producing efficiency of the *Bt*. gene gradually goes down and sometimes in the later stages clear cut differences are not observed in the *Bt*. and non-*Bt*. cotton with regard to attack of bollworms. However, in 90 to 120 days the major portion of boll formation takes place and there is little effect on the productivity of cotton.

2. Ineffective against Sucking Pests

The *Bt.* cotton has been developed to confer resistance against bollworms for which resistant genotypes are not available in the cultivated as well as wild germplasm of cotton the world over. Bollworms are considered as enemy number one of cotton crop. *Bt.* cotton provides effective control of three types of bollworms. However, it is not effective in controlling sucking pests. Hence, efforts are needed to develop transgenic cotton with resistance to bollworm as well as sucking pests.

3. Effect on Insecticide Industries

The development of *Bt.* cotton is going to reduce the pesticide use significantly. The reduction in the pesticide use has been reported to the tune of 38 per cent. This will have adverse effect on the insecticide industries. In cotton crop, now insecticides are mainly required for the control of sucking pests. Earlier about 45 per cent of total pesticides on global level and 50 per cent in India was applied to cotton crop alone.

4. Effect on Employment

When there will be very limited demand for the insecticides, many of the insecticide industries will reduce production of insecticides. As a result, there will be reduction in the strength of persons engaged in these industries. This will have adverse effect on the employment.

5. Promotes Malpractices

Comparatively higher price of *Bt.* cotton seed promotes malpractices such as mixing of non-Bt. seed with *Bt.* and sale of F_2/F_3 seeds. This will have adverse effect on cotton production and thereby on the economic condition of the farmer using such seed. Steps should be taken up by the Government to curb such unhealthy practices.

FUTURE THRUSTS

The genetic resistance is the cheapest and the most efficient method of protecting crop plants from pests. The *Bt.* transgenic cotton with inbuilt genetic resistance to bollworms helps in protection of natural enemies of insect pests *i.e.* predators and parasites. It also helps in reducing the cost of cultivation by reducing the use of pesticides. Moreover, it reduces environmental pollution and health hazards caused by pesticide use. Transgenic cottons with *Bt.* endotoxin protein reduces expenditure on insecticides and create eco-friendly environment without reduction in yield. The future research on *Bt.* transgenic cottons needs to be directed towards following thrust areas:

1. Diversification of *Bt.* Sources

Through widespread cultivation of *Bt.* transgenic cotton, the main risk is development of insect resistance against *Bt.* toxin. Hence, multiple sources of resistance should be identified and used in developing bollworm and herbicide resistant *Bt.* transgenic cottons to avoid the risk of developing insect resistance and herbicide resistant weeds.

Besides *Bt.* gene, several other genes can be used in future for developing resistant genotypes of cotton to various insects. Some examples are given below:

(i) Cholesterol oxidaze gene from *Streptomycetes* fungus can be used for developing boll weevil resistant genotypes.

(ii) The spider and scorpion venom genes can also be used for developing insect resistant genotypes of cotton.

(iii) The *Helicoverpa armigera* stunt virus contains three genes which attack midgut of *Heliothis* and ceases its feeding.

(iv) Protease inhibitor gene from cowpea, soybean and rhizomes of African Taro may be tried for development of transgenic cotton.

2. Use of Male Sterility

Several transgenic *Bt.* cotton hybrids have been released for commercial cultivation in India. However, all these hybrids have been developed without the use of male sterility. Use of male sterility eliminate the process of emasculation and thus help in further reducing the price of *Bt.* cotton. Hence, there is need to develop male sterility based *Bt.* cotton hybrids.

3. Drought Resistant *Bt.* Cotton

Prior to the introduction of *Bt.* cotton, there were two major problems related to cotton in India. The first problem was yield losses due to bollworms which was common in all the three cotton growing zones. Development of *Bt.* cotton has provided solution to the problem of bollworms. The second major problem is of moisture stress. In India, about 65 per cent of the cotton crop is grown under rain-fed conditions. The problem of soil moisture stress is confined to central and south zones where cotton crop suffers due to moisture stress at one or the other stage resulting in heavy yield losses. Hence, there is need to develop drought resistant *Bt.* cotton for such areas. Osmotin gene, dehydrin gene and some other drought resistant genes available can be used for this purpose.

4. Salinity Resistant *Bt.*

In some areas of Gujarat, Andhra Pradesh and Tamil Nadu where cotton is grown, there is problem of soil salinity. Cotton crop also suffers from abiotic stresses such as drought and salinity. There is need to develop *Bt.* transgenic cottons with resistance to salinity conditions.

5. *Bt.* Varieties

In case of hybrids, the farmer has to purchase fresh seed every year. Only one *Bt.* Variety 'Bikaneri Narma' has been developed jointly by Central Institute for Cotton research, Nagpur and University of Agricultural Sciences, Dharwad. Efforts are needed both by public and private sectors to develop many more such varieties to provide wider choice to the farmers. The seed of such varieties can be used by the farmers for 3-4 years.

5. Diploid *Bt.* Hybrids

Diploid cottons cover about 15 per cent of cotton area in India. Hence, there is need to develop transgenic *Bt.* varieties and hybrids of diploid cotton.

6. Cotton Leaf Curl Resistance

There is problem of cotton leaf curl virus in the north zone. Besides use of donor lines from germplasm, biotechnology can also be used for development of *Bt.* hybrids resistant to this disease.

7. Improvement in Quality

Cotton is a fibre, oil and protein yielding crop. There is need to improve the quality of proteins and oil through genetic engineering besides fibre quality improvement. There is need to develop varieties and tetraploid hybrids with high fibre strength suitable for high speed spinning[jet and rotor spinning]. Such cotton will also fetch premium price in the International market.

8. *Bt.* Barbadense

Egyptian cotton [*Gossypium barbadense*] cotton is cultivated in small area [>1 per cent] in Tamil Nadu and Andhra Pradesh. This is a quality cotton which belongs to extra long staple group and used for manufacturing superfine clothes. Hence, there is need to develop high yielding *Bt.* variety of this species better than the existing variety Suvin.

9. Development of *Bt.* hybrids and varieties suitable for machine picking.

10. Development of short duration, short stature and compact *Bt.* cultivars and hybrids in upland cotton to achieve quantum jump in the productivity by adopting closer spacing.

11. Development of *Bt.* varieties and hybrids with high ginning outturn [more than 40 per cent] both in diploid and tetraploid cottons.

SUMMARY

Genetically engineered organism are called transgenics. *Bt.* cotton refers to transgenic cotton which contains endotoxin protein inducing gene from soil bacterium *Bacillus thuringiensis.*

We do not have any source of resistance to the bollworms in the germplasm of cotton including wild species and wild relatives the world over. Moreover, about 10 per cent of insecticides on global basis and 45 per cent in India are used for control of insects in cotton crop alone. There are several adverse effects of Insecticides. The main adverse effects include killing of (i) parasites and predators, (ii) beneficial insects, (iii) beneficial micro-organisms, (iv) environmental pollution, (v) increase in the cost of cultivation, (vi) development of resistance in the insects, and (vii) adverse effects on animal and human health.

Some *Bt.* genes resistant to bollworms were identified in Japan, USA and India between 1990 and 2002. A list of important genes identified by different Seed

Companies/public sector organizations so far is presented. The Chronology of the development of *Bt.* Cotton in India is briefly presented.

The transgenic cotton can be divided into different categories on the basis of (i) resistance, (ii) genes involved, (iii) method of development, and (iv) material used.

In genetic engineering, an event is the insertion of a particular piece of foreign DNA into the chromosome of the recipient crop plant. All *Bt.* cotton genotypes till 2009 [hybrids + Variety] have been developed using six different events, *viz.,* by Monsanto event 531, Monsanto event 15985, IIT Kharagpur event 1, Chinese GFM event, Dharwad event and Metahelix event [02].

In cotton, two significant achievements have been made so far through application of biotechnology. These are: (1) bollworm resistance, and (2) herbicide resistance. These have been amply discussed. Limitations and future thrusts of *Bt.* cotton have been discussed.

QUESTIONS

1. Define *Bt.* Cotton and describe its main features.

2. Discuss briefly advantages of transgenic technology.

3. Describe briefly progress of transgenic cotton breeding in India.

4. Give a brief account of practical achievements of transgenic cotton breeding in North Zone.

5. Give a brief account of practical achievements of transgenic cotton breeding in Central Zone.

6. Give a brief account of practical achievements of transgenic cotton breeding in South Zone.

7. Define an event. Describe briefly different events used in cotton breeding in India.

8. Discuss briefly future thrusts of transgenic cotton breeding in India.

9. Explain briefly limitations of transgenic cotton breeding.

10. Describe briefly various types of transgenic cotton hybrids.

Chapter 10

DUS Testing

INTRODUCTION

Evaluation of plant varieties in terms of distinctiveness, uniformity and stability is referred to as DUS testing. The testing of DUS characters is useful in the following five main ways *viz.*, (i) for identification of plant varieties, (ii) registration and protection of plant varieties under Protection of Plant Varieties and Farmers Rights Act 2001 (iii) varietal information system, and (iv) classification of varieties in to different groups, and (v) creating data base for plant genetic resources.

Government of India has passed an Act entitled Protection of Plant Varieties and Farmers' Rights Act, 2001. The main mandates of the Act are: (i) registration of plant varieties, (ii) characterization, cataloguing and documentation of registered varieties, (iii) ensuring that seeds of all registered varieties are made available to farmers, and (iv) maintenance of National Register of Plant variety.

FARMERS' RIGHTS

Provision of Farmers' Rights in the Act is unique to the Indian system of Plant Variety Protection. It provides rights to the farmers to:

(i) Register their varieties.

(ii) Share benefit for use of biodiversity conserved by farming community.

(iii) Save, use, sow, re-sow, exchange, share or sell farm produce including seed of registered variety (not branded).

(iv) Claim compensation for under performance from the promised level.

(v) Seek their consent when farmer variety is used to develop an essentially derived variety (EDV).

BREEDERS' RIGHTS

Breeders' rights allow researchers to use protected variety for the purpose of research and breeding new varieties and for commercializing them with out securing prior permission from the PBR holder of the initial variety.

Compulsory Denomination

Any protected variety must be given a single and distinct denomination in accordance with the regulations. Any person who markets the produce of this protected variety must use the same denomination even if the period of protection is expired which is again a valuable part of Breeders' rights.

TYPES OF VARIETIES

In connection with plant variety protection (PVP) Act, various terms such as extant variety, candidate variety, reference variety, example variety and farmers variety are frequently used. Hence, knowledge of these terms is essential. These are defined below:

Extant Variety

All Released, notified and unprotected varieties come under this group.

Candidate Variety

A variety to be registered under Plant Variety Protection Act is referred to as candidate variety.

Reference Variety

All released and notified extant varieties of common knowledge which are in seed production chain.

Example Variety

A variety that is used for comparison for a particular character is called example variety.

Farmers' Variety

A variety that has been developed by a farmer and used for commercial cultivation for several years is called farmers variety.

REQUIREMENTS FOR PROTECTION

The PPV and FR Act, 2001 provides that protection shall only be granted after examination of the variety for Distinctness, Uniformity and Stability and Novelty. The first three requirements were as per UPOV Act, 1978 and are known as DUS. The fourth one was included in UPOV Act, 1991 and known as NDUS.

(i) Distinctiveness

The new variety must be clearly distinguishable by one or more characters from any other variety whose existence is a matter of common knowledge at the time when the protection is applied for.

(ii) Uniformity

The variety should be sufficiently uniform in its relevant characteristics. Relevant characteristics include all those used as a basis for distinctiveness or included in the variety description established at the date of grant of protection of that variety.

(iii) Stability

The variety is deemed to be stable if its relevant characteristics remain unchanged after repeated propagation.

(iv) Novelty

It refers to newness of a variety. The variety should be new one and should not have been commercially cultivated for more than one year before granting protection under PVP Act.

Period of Protection

The period of protection varies with plant species. For field crops, the maximum period of protection is 15 years, whereas for forest trees, fruit trees, ornamental trees, shrubs and vines it is 18 years.

HOW TO CLAIM PBR

Registration of any new plant variety under the PPV FR is voluntary. It consists of following steps.

1. An application in the prescribed form varietal description bringing out Distinctness, Uniformity, Stability and Novelty.
2. Complete passport data of parental lines, their geographical origin;
3. An affidavit declaring absence of terminator gene sequence in the variety was lawfully acquired.
4. Specified fee for registration and conducting DUS testing; prescribed quantity of seed of the variety and parental lines (for hybrids) in such a manner and conduct that viability and quality shall remain unaltered.

DUS TESTING IN COTTON

The details of DUS testing in cotton are briefly discussed as follows:

Material Required

The minimum quantity of seeds to be supplied by the applicant is 2 kg for varieties, hybrids and parental line in cotton which has to be provided in one submission. The seeds should meet the minimum requirement of germination (> 80 per cent), moisture content (< 10 per cent) and physical purity (98 per cent for varieties, hybrids and their parents), highest genetic purity. The seed should not have undergone any treatment unless the authority allows it.

Conduct of Test

The minimum duration of tests should normally be two independent similar growing seasons at two locations. The test conditions should ensure normal growth of the plant. The plot design is as follows:

Number of rows: 12

Row length: 6 m

Row to row distance: 90 cm

Plant to plant distance: 60cm

No: of replications: 3

Method of Observation

The characteristics described in the table of characteristics should be used for the testing of varieties, inbred lines and hybrids. The observations should be made on 40 plants which should be divided among 4 replications (10 plants/replication). For uniformity, testing is done by visual assessment of a group of plants in single observation. The number of aberrant plants should not exceed 8 in 150. All leaf characteristics should be observed on 4th leaf from the top.

DUS Testing Centres

In cotton, four DUS testing centres, *viz.*, Nagpur, Coimbatore, Dharwad and Hisar have been identified. The Nagpur and Hisar centres will deal with testing of *G. hirsutum* and *G. arboreum* species. Coimbatore centre will test *G. hirsutum* and *Gossypium barbadense* species and Dharwad centre will test varieties of *G. hirsutum* and *G. herbaceum*.

CHARACTERS FOR DUS TESTING

As per National DUS Test Guidelines, in cotton, 41 characters have been decided for DUS testing. These traits are related to seedling (1), plant (3), stem (2), leaf (9), bract (2), flower (6), boll (8), fibre (7) and seed (3) and include both polygenic/ quantitative (governed by many genes) as well as oligogenic/qualitative traits (governed by very few genes).

The characteristics are again classified into three groups, *viz.*, (i) Grouping characters, and (ii) Essential characters and (iii) Additional characters.

(i) Grouping Characters

These are highly heritable characters which are generally not found to vary within variety under any condition and hence suitable for grouping of varieties into different classes. Grouping characters include leaf shape, petal color, boll shape and fibre length [Table 10.1].

TABLE 10.1: Characters for Grouping in Cotton.

1.	Leaf shape	Palmate (normal), semi-digitate (semi-okra), digitate (okra) and laceolate (super okra).
2.	Petal color	White, cream, yellow, pink, red and dark red.
3.	Boll shape	Round, oval and elliptic.
4.	Fibre length	Very short, short, medium, long and extra-long.

(ii) Essential Characters

These are trait which have been showing the most variation among the varieties and are necessarily to be recorded and included in the variety description. In cotton, there are 20 essential characteristics as follows which also includes the grouping characters [Table 10.2].

TABLE 10.2: Essential Characters for DUS Testing in Cotton

1.	Leaf shape	Palmate, Semi–digitate, digitate and super-okra.
2.	Leaf size (width at maximum point)	Small, medium and large.
3.	Leaf color	Light green, green, light red and dark red.
4.	Leaf : Pubescence	Absent, medium and strong.
5.	Leaf nectaries	Absent and present.
6.	Bract type	Normal and frego.
7.	Petal colour	White, cream, yellow pink, red and bi-color.
8.	Petal spot	Absent and present.
9.	Position of stigma	Embedded and exerted.
10.	Anther color	White, cream, yellow and purple.
11.	Boll size	Small, medium and large.
12.	Boll shape	Round, oval, elliptical.
13.	Boll surface	Smooth and pitted.
14.	Boll opening	Open, semi-open and close.
15.	Fibre length	Very short, short, medium, long and extra-long.
16.	Fibre color	White, cream, brown and green.
17.	Fibre strength	Weak, medium and strong.
18.	Ginning per cent	Low, medium, high and very high.
19.	Seed : Density of fuzz	Naked, semi-fuzzy and fuzzy.
20.	Seed: Fuzz color	White, grey, brown and green.

(iii) Optional Characters

There are 21 characters that may or may not be recorded as per requirement and is given in the table of characteristics. List of optional characters along with their categories is presented in Table 10.3.

TABLE 10.3: Optional Characters for DUS Testing in Cotton

1.	Hypocotyl: pigmentation	Absent and present.
2.	Plant: time of flowering (50 per cent of plants with at least one open flower)	Early (<45 days), medium (45-60 days) and late (>60 days).
3.	Plant: stem pigmentation	Absent and present.
4.	Plant: stem hairiness	Absent, sparse, medium and strong.
5.	Leaf: lobe number	One, three, five and seven.
6.	Leaf: appearance	Cup and flat.
7.	Leaf: gossypol glands	Absent and present.
8.	Leaf: petiole pigmentation	Absent and present.
9.	Bract: number of serration	Few, medium and many.
10.	Flower: Male sterility	Absent and present.
11.	Flower: filament pigmentation	Absent and present.
12.	Boll: bearing habit	Solitary and cluster.
13.	Boll: color	Green and red.
14.	Boll: prominence of tip	Blunt and pointed.
15.	Boll: weight of seed cotton/boll	Small (<3.0 g), medium (3.1-5.0 g.) and large (>5.0 g).
16.	Plant: growth habit	Determinate and indeterminate.
17.	Plant: height	Very short (<61 cm), short (61-90 cm), medium (91-120 cm), tall (121-150 cm) and very tall (>150 cm).
18.	Seed: size (100 seed wt.)	Very small (<5.1 g), small (5.1-7.0 g), medium (7.1-9.0 g), bold (9.1 – 11.0 g) and very bold (> 11 g).
19.	Fibre: fineness (micronaire value)	Very fine (<3.0), fine (3.0-3.9), medium (4.0-4.9), coarse (5.0-5.9) and very coarse (> 5.9).
20.	Fibre: uniformity	Poor (<40), average (40-45) and good (>45).
21.	Fibre: maturity (per cent)	Poor (<70), average (70-80) and good (>80).

There are three more additional characters which are not yet included in the table of characteristics but could be observed as an additional character if sufficient variability could not be obtained from the above characteristics. These are vein/ sepal pigmentation, bract pigmentation and petiole pigmentation.

SUMMARY

DUS testing refers to evaluation of varieties in terms of distinctiveness, uniformity and stability. DUS testing is useful for identification of varieties, registration of varieties under PPV and FR Act 2001, varietal information system and creating data base for plant genetic resources. There are four characters which are used for grouping. Such characters include leaf shape, petal colour, boll shape and fibre length. These are highly heritable characters that are not influenced by environmental effects. For DUS testing, characters are classified in to two groups, *viz.*, (i) essential characters [20 characters] and (ii) optional characters [21 characters]. Essential characters have to be recorded, whereas optional characters may or may not be recorded.

In cotton, there are four DUS testing centres, *viz.*, CICR, Nagpur, CICR (RS), Coimbatore, UAS Dharwad and HAU, Hisar. Nagpur and Hisar centres will carry out testing of *G. hirsutum* and *G. arboreum*; Coimbatore for *G. hirsutum* and *Gossypium barbadense*; and Dharwad for *G. hirsutum* and *G. herbaceum*. The test has to be conducted at least at two locations for two normal seasons. Each test should be based on minimum sample of 150 plants which have to be divided in to four replications.

QUESTIONS

1. Define DUS. What are the requirements for DUS testing?

2. Discuss briefly advantages of DUS testing.

3. Describe briefly Plant Breeders Exemptions and Farmers' rights.

4. Give a brief account of various types of varieties in relation to DUS testing

5. Write short notes on the following:
   (i) Period of protection for trees and vines
   (ii) Period of protection of field crops
   (iii) Essential characters for DUS testing
   (iv) Optional characters for DUS testing

6. Define the following terms:
   (i) Novelty
   (ii) Distinctiveness
   (ii) Uniformity
   (iv) Stability

7. Write short notes on the following:
   (i) Candidate variety
   (ii) Extant variety
   (ii) Example variety
   (iv) Reference variety

8. Write short notes on the following:
   (i) Breeders exemption
   (ii) Farmers rights
   (ii) Novelty
   (iv) Farmers' variety

9. Write short notes on the following:
   (i) Requirements for DUS testing
   (ii) Advantages of DUS testing

Chapter 11

Cotton Seed as Source of Edible Oil

INTRODUCTION

Cotton is a multipurpose crop. It is a fiber, oil and protein yielding crop. However, cotton crop is primarily cultivated for fibre. The cotton seed which is a byproduct is an important source of edible oil. Cotton seed is the second largest source of edible vegetable oil in the world. Refined cotton seed oil is one of the best edible oils. It is used in most parts of the world including U.S.A., Uzbekistan, China and Middle East. Now, cotton seed oil is also used for edible purposes in India. Hence, genetic improvement in cotton seed oil content without reduction in the lint yield will be an added advantage. On global level, cotton seed is the third highest source of edible vegetable oil after soybean and Rapeseed [Table 11.1]. In India, cotton seed is the second highest source of edible vegetable oil after soybean. Rapeseed/mustard and groundnut occupy third and fourth position respectively [Table 11.2].

TABLE 11.1: World Production of Major Oilseeds [In million tones]

Sl.No.	Oilseed Crops	2013-2014	2014-2015	2015-2016	2016-2017
1	Soybean	283.7	314.5	320.1	345.9
2	Rapeseed	71.3	70.9	68.0	68.5
3	Cottonseed	45.0	45.6	37.3	42.3
4	Groundnuts	38.9	37.9	40.7	45.4
5	Sunflower	42.3	40.7	39.1	38.9
6	Palm-Kernel	14.6	15.3	16.3	17.0
7	Copra	5.6	5.7	5.5	5.4
8	**Total**	**501.4**	**530.6**	**528.0**	**563.4**

TABLE 11.2: Production of Major Oilseeds in India [In Lakh tones]

Sl.No.	Oilseed Crops	2009-2010	2015-2016
1	Cotton seed	86.5	115.0
2	Soybean	85.0	125.0
3	Rapeseed	64.2	90.0
4	Groundnuts	35.8	64.0
5	Sunflower	9.9	18.0
6	Castor	9.3	7.1
7	Sesame	7.6	8.2
8	Safflower	1.5	1.1
9	Linseed	1.6	1.9
10	Niger	0.8	1.1
	Total	**428.4**	**431.4**

QUALITY OF COTTON SEED OIL

Cotton seed oil possesses several good qualities as discussed here as under:

(i) Heart Healthy

The cotton seed oil is versatile for edible purposes. It is heart healthy oil which is free from cholesterol and contains low level of saturated fatty acids and trans-fatty acids.

(ii) High Smoke Point

It has high smoke point and hence can be used at very high temperature for frying of potato chips and other snacks.

(iii) Rich in Essential Fatty Acids

Cotton seed oil is rich in essential fatty acids such as myristic, palmitic, palmitoleic, stearic, oleic and linoleic acids. Linoleic acid, which is the most important one, is present to the extent of 51 per cent (Shaikh *et al.,* 1996). The deficiency of above acids, leads to narrowing of arteries causing reduced blood supply to the heart.

(iv) Uses

Presently, refined cotton seed oil is widely used for edible purposes. It is also used for making hydrogenated oils.

(v) Other Qualities

The cotton seed oil is highly stable and has longer self-life due to presence of tocopherol which acts as an antioxidant. The crude oil is more stable due to presence of gossypol along with tocopherol. Moreover, the cotton seed oil has lesser tendency to undergo flavor reversion.

AVAILABILITY OF COTTON SEED

In India, the current availability of cotton seed is ranging from 91.5 to 115 lakh tones per annum. Out of this, 25 lakh tones is used for sowing and cattle feed. For crushing purposes, about 86.5 to 90.0 lakh tones seed is available annually. Annually about 10 lakh tones cotton seed oil is obtained [Table 11.3].

TABLE 11.3: Annual Availability of Cotton Seed (Lakh Tones)

Sl.No.	Particulars	2009-2010	2015-2016
1	Total production of cotton seed	91.5	115.0
2	Used for sowing and cattle feed	5.0	25.0
3	Marketable surplus	86.5	90.0
4	Production of washed cotton seed oil [12.5 per cent recovery]	10.8	10.0

GENETIC RESEARCH

The genetic research on cotton seed oil has been limited to pattern of variability for seed oil content and fatty acid contents, gene action, level of heterosis, correlations, heritability, maternal effects, *etc.*

(i) Pattern of Variability

The pattern of variability for seed oil content has been studied in all the four cultivated species. The extent of variability has been reported to be higher in tetraploid cottons than diploid. The extent of variability observed in the germplasm of four cultivated species indicates ample scope for genetical improvement of cotton seed oil through hybridization followed by directional selection.

In *G. hirsutum,* the seed oil content ranges from 10.2 to 29.8 per cent. The high seed oil germplasm lines (26-29.8 per cent) included Acala 5-1, ALPPO-1 x Uganda 1-121-17-174, Acala Q-6-1, ALPPO-1 Empire Glandless, B-58-1290, Acala 1517-BR-2, X-82, 5-44 and 561. The high seed oil *hirsutum* varieties (23.7-26.1 per cent) included LH 1556, F 2030, F 414, H 655 C, SGNR 10, RS 810, LRK 516, Indore 2, MCU 2 and DS 59. High seed oil Intra-*hirsutum* hybrids included CSHH 238, LHH 144 and Hybrid 4.

In *Gossypium barbadense,* the seed oil content ranges from 14.0-25.8 per cent. The high seed oil germplasm lines (23.3-25.8 per cent) included EC 97635, Egypt-1, EC 98252, EC 97634, St.Kitts, Marrad, 5230 and K-4831-56015. In this species, there are only two released varieties *viz.,* Suvin and Sujata, which contains 24.9 and 25.8 seed oil per cent respectively.

In *G. arboreum,* the seed oil content ranges from 12.5-25.8 per cent. The high seed oil germplasm lines (23.0-24.5 per cent) included H 474, 18 B, Desi 72, Indicum-1290, Verum 262, W-31-O-B, 57-94, 091, Sarguja NL-WF, X-89 and Western Bani. The high seed oil *arboreum* cultivars included RG 8, Y1, AK 235 and AKA 5. In these cultivars the seed oil content was 24, 23.2, 22.6 and 22.4 per cent respectively.

In *G. herbaceum*, the seed oil content ranges from 13.5-20.4 per cent. There was only one germplasm line *viz.*, LS Early 2 which was having more than 20 per cent seed oil content. Two released cultivars *viz.*, Digvijay and Sujay recorded seed oil content of 21.4 and 22.4 per cent respectively.

(ii) Variability for Fatty Acid Contents

The variability for different saturated and unsaturated fatty acids has been studied by various workers. The extent of variability observed in the germplasm of four cultivated species indicated ample scope for genetic improvement of unsaturated fatty acids *i.e.* linoleic acid and oleic acid. The range of variability for saturated (palmitic and stearic) fatty acids and unsaturated fatty acids (oleic and linoleic acids) is presented in Table 11.4.

TABLE 11.4: Variability for Fatty Acid Content

Sl.No.	Species	Palmitic	Stearic	Oleic	Linoleic
1	G. hirsutum	23.1-28.0	2.4-3.4	14.7-20.9	47.6-55.4
2	G. barbadense	24.4-25.4	2.6-3.0	18.7-19.7	50.0-51.7
3	G. arboreum	18.9-21.2	1.1-3.4	16.5-30.7	30.3-59.3
4	G. herbaceum	20.5-23.4	3.2-4.2	17.5-20.8	51.3-55.1

In *G. hirsutum*, the linoleic acid content ranged from 47.6-55.4 per cent and oleic acid content from 14.7-20.9 per cent. The germplasm lines with high linoleic acid content included B-56-181, DCI 116, MA-7, DCI 122, DHY 286, J-34, G.Cot 100, ACHH-703F, Badnawar and 6088. The superior germplasm lines for oleic acid content included M-26 cc, PKV 804, MCU 5, M 19 cc, LH 900, PKV 802, 150-3-1, Indore-2, SS-265 and 21.

In *Gossypium barbadense*, the linoleic acid content ranged from 50-51.7 per cent and oleic acid content from 18.7-19.7 per cent. In *G. arboreum*, the linoleic acid content ranged from 30.3-59.3 per cent and oleic acid content from 16.5-30.7 per cent. The superior germplasm lines for linoleic acid content included Malvi-20, AKH 235, Malsoni, Coconada white and Jarilla. The superior lines for oleic acid content were Gaorani-2, Gaorani-CB-5, G-21, Malsoni and Malvi-2.

In the four cultivated species of cotton, the fatty acid contents ranged from 14 to 26 per cent for palmitic acid, from 0.4-4.7 per cent for stearic acid, from 16-28 per cent for oleic acid and 40-56 per cent for linoleic acid. The oleic and linoleic acids are unsaturated fatty acids whereas palmitic and stearic acids are saturated fatty acids.

(iii) Gene Action

Gene action for seed oil content has been studied by some workers. Three types of results have been reported. In some studies, seed oil content was controlled by additive genes, while results of another study revealed that seed oil content is primarily governed by non-additive gene effects. In one study, both additive and non-additive gene effects were reported important for seed oil content. However, the dominant component was higher than additive component. Thus, both additive and non-additive gene effects are important for the expression of seed oil content.

(iv) Heterosis

A very few reports are available on heterosis of seed oil content. The level of heterosis has been reported over mid-parent, better parent and commercial cultivar. The high level of heterosis for seed oil content was generally observed in the crosses involving geographically or genetically distant parental lines (Table 11.5).

TABLE 11.5: Level of Heterosis Observed for Seed Oil Content (Maximum)

Sl.No.	Type of Crosses	Heterosis over	Heterosis (per cent)
1	H x H Crosses (204)	Mid parent	27.6
		Better parent	21.0
		Check variety	22.5
2	A x A crosses (51)	Mid parent	20.1
		Better parent	14.8
		Check variety	29.6

(v) Combining Ability and Heritability

Combining ability is an important measure of studying gene action. Combining ability for seed oil content has been studied by various workers. Both GCA and SCA variances were found important in expression of seed oil content. This suggests that seed oil content is governed by both additive and non-additive gene action. In a study, heritability estimates for oil content ranged from 40 per cent to 53 per cent.

(vi) Correlations

In cotton, very few reports are available on correlation of seed oil content with various economic characters. The correlation of seed oil content was found to be positive with seed size, okra leaf, non-fruiting branches, earliness, linoleic acid and fibre length; and negative with protein content, palmitic acid, stearic acid and oleic acid contents.

(vii) Maternal Effects

In cotton, seed and embryo size are determined predominantly by maternal parent. Various workers have studied maternal effects as source of variation for seed oil content in cotton. It was reported that maternal genetic effects are more important than direct effect for seed oil percentage. Hence varieties with high seed oil content may be developed by direct selection based on maternal plants.

SOURCES OF HIGH SEED OIL CONTENT

In cotton, there are four important sources of high seed oil content, *viz.*, cultivated varieties, germplasm collections, wild species, and induced mutations. The first two sources can be easily used for genetic improvement of seed oil content.

(i) Cultivated Varieties

Cultivated varieties are very good sources of seed oil content. All the old and currently cultivated cultivars should be evaluated for identification of high seed oil content.

(ii) Plant Genetic Resources

Germplasm collections are important sources for improvement of seed oil content. The global gene pool of all the four cultivated species is maintained at the Central Institute for Cotton Research, Nagpur. Part of this germplasm has been evaluated for seed oil content and high seed oil lines have been identified which can be utilized in breeding programs for genetic improvement of seed oil content.

(iii) Wild Species

Wild species are potential sources of high seed oil content. Some wild species possess high seed oil content. However, utilization of wild species in the hybridization programs poses several problems such as cross incompatibility, hybrid inviability, hybrid sterility, *etc.* Hence, wild species are used only when good source of seed oil content is not found within the cultivated species.

(iv) Induced Mutations

The seed oil content can also be improved through the use of induced mutations. However, this source is also rarely used for genetic improvement of seed oil content.

BREEDING METHODS

Breeding methods for genetic improvement of seed oil content are the same as for other agronomic characters. Breeding methods, which can be used, for improvement of seed oil content include backcross, pedigree method, recurrent selection and mutation breeding. Pedigree selection proved effective in improving seed oil content in upland cotton without affecting other agronomic characters. Disruptive selection was found effective in improving the seed oil content in upland cotton. Harland (1949) reported wide range of variability (21.8-29.2 per cent) for seed oil content in 7 lines of Egyptian cotton and indicated scope for genetic improvement seed oil content by 7 per cent in commercial cultivars. Mutation breeding has been used and found effective in improving seed oil content in upland cotton.

FACTORS AFFECTING SEED OIL CONTENT

The estimate of seed oil content varies depending upon four main factors, *viz.,* genetic material, crop harvest, method of estimation, and cultivation practices. These are discussed as follows:

(i) Genetic Material

The seed oil content varies depending upon cotton species. Generally, the seed oil content is higher in tetraploid cottons than diploid ones. Genotypic differences exist even among the genotypes of a species and also among the races of a species.

(ii) Crop Harvest

The seed oil content is generally higher in the seeds obtained from the first picking. Higher seed oil content has been reported in the seeds obtained from the bolls, which developed in first 3-4 weeks after flowering.

(iii) Method of Estimation

The seed oil estimates also vary depending upon the method used for seed oil estimation. Generally, the seed oil estimates are low with crushing or pressing method. The last method gives about 1-2 per cent higher estimates than ether solvent method.

(iv) Cultivation Practices

The type of cultivation also affects seed oil content. Generally, seed oil content is more in irrigated crop than rain-fed one. Climatic factors like rainfall and temperature, management factors like irrigation and mineral nutrition, biotic and abiotic stresses, and interaction of all these factors with genetic constitution of a line are known to affect the cottonseed oil content and quality.

OIL ESTIMATION TECHNIQUES

There are three methods, which are used, for estimation of seed oil content in the breeding materials. These are crushing or pressing method, solvent extraction method and non-destructive method using NMR.

(i) Crushing Method

This method of oil extraction is very old. In this method, the oil is extracted through hydraulic pressing or screw pressing. This method is used for extracting seed oil in large quantity. Now small, medium and large crushing machines are available. This method has two main drawbacks. Firstly, it cannot be used for extraction of seed oil from small seed samples obtained from single plants. Secondly, the recovery of oil in this method is very low. Hence, this method is not used for scientific investigations.

(ii) Solvent Extraction Method

This method is widely used in research laboratories for scientific studies related to seed oil content. In this method, first the seed is ground into flour, which is used for oil extraction. Petroleum ether is used as solvent. The flour is mixed with ether thoroughly and then filtered. The filtrate is heated in the Soxlet apparatus for 6-8 hours and condensed liquid is collected in the receiver flask. Then the receiver flask is kept in oven at 100-105°C for evaporation of petroleum ether. The weight of the extract is recorded and calculated in percentage. This method is very much accurate but time consuming.

(iii) Nuclear Magnetic Resonance (NMR)

This technique is a rapid and non-destructive method of seed oil analysis. This method is quite simple, accurate and very fast. By this method, 300-400 samples can be easily analyzed per day. The pure oil of cotton is used for calibration. The

NMR is of two types *viz.*, simple NMR and pulsed NMR. In simple NMR, drying and weighing of samples is required, while in pulsed NMR, oil analysis can be done without drying and weighing of sample. This method simultaneously provides estimates of seed oil and seed moisture. Analysis of a sample takes a few seconds (8-32 seconds). The results are printed by the attached printer directly in percentage.

In India, the average recovery of cotton seed oil by crushing and pressing method is 10-12 per cent whereas in U.S.A., the oil recovery is upto 20 per cent. The potentiality of oil recovery exists upto 25 per cent. Through traditional processing, the seed oil content remains upto 7 per cent in the cake whereas by scientific processing, the seed oil content remains less than 1 per cent in the cake.

PRACTICAL ACHIEVEMENTS

In upland cotton, seed oil improvement has been achieved through disruptive selection, pedigree method and mutation breeding. These are discussed as follows:

(i) Disruptive Selection

In upland cotton, lines having 1.8 to 5.7 per cent higher seed oil content than parents have been developed through disruptive selection.

(ii) Pedigree Selection

In two crosses of upland cotton, the seed oil content increased from 20.5 to 24.2 per cent and 20 to 22 per cent respectively after seven years of pedigree selection.

(iii) Mutation

A dark green foliage mutant obtained by ionizing radiation had 25 per cent higher seed oil content than parent variety. A mutant with increased fatty acid content was isolated from progeny raised after gamma irradiation. In upland cotton, lines having 1.4 to 2.5 per cent higher seed oil than parent line were isolated. Mutant lines with enhanced oleic acid content were obtained through treatment of floral bud with 0.05 per cent solution of Ethylene imine.

FUTURE THRUSTS

In cotton, several studies have been made on seed oil content related to variability, heterosis, combining ability and gene action, correlations *etc*. The future research work for improvement of cottonseed oil needs to be directed towards the following thrust areas:

1. Improvement in Seed Oil Content

The average seed oil content in the presently cultivated varieties and hybrids is about 20-22 per cent. The seed oil percent in germplasm has been recorded upto 27 per cent. Thus, there is a gap of 5-6 per cent, which could be realized through appropriate breeding techniques.

2. Development of Glandless Cultivars

All presently cultivated cotton varieties and hybrids are glanded. The oil extracted from such varieties requires refinement to make it suitable for human consumption. There is ample scope to develop high seed oil lines with glandless trait in the seed and glanded vegetative parts to bring down the cost of cotton seed oil.

3. Improvement in Oleic Acid Content

In the presently available cotton hybrids and cultivars, the amount of oleic acid needs further improvement.

4. Reduction in Hull Content

The proportion of hull in the presently available cultivars and hybrids is quite high. There is ample scope to reduce the hull content and increase the kernel portion in the cotton genotypes.

5. Use of Modern Techniques

The modern breeding techniques such as transgenic development, molecular breeding and marker aided selection may be rewarding in developing lines with high seed oil content.

6. Use of Foreign Gene

There is ample scope to incorporate foreign gene through biotechnology in glandless lines to enhance seed oil quality and productivity.

SUMMARY

Cotton seed oil is the second largest source of edible oil in India. Refined cotton seed oil is one of the best edible oils and is used in USA, Uzbekistan, China and middle-east for human consumption. Genetic improvement in seed oil content without bringing reduction in the lint yield will be an added advantage.

Genetic studies on seed oil content have been limited to variability, gene action, heterosis combining ability, heritability and maternal effects. Exploitable extent of variability has been observed for seed oil content in all the four cultivated species. There are few reports on gene action of this character. Both additive and non-additive gene actions were found important for the expression of seed oil content. Moderate heterosis for seed oil content has been reported both in upland and *G. arboreum* cottons. There are four important sources of high seed oil content, *viz.*, cultivated varieties, germplasm collections, wild species and induced mutations.

Breeding methods, which are used, for genetic improvement of seed oil content are the same used for genetic improvement of other polygenic traits. Important methods include backcross, pedigree method, recurrent selection, disruptive selection and mutation breeding. Several genotypes of cotton with high seed oil content have been developed by these methods. There are three methods, which are used for estimation of seed oil content, *viz.*, crushing or pressing method, solvent extraction method and non-destructive NMR method. The last two methods are used for evaluation of cotton seed samples for seed oil content. The estimates of seed oil content are affected by various factors such as genetic material, crop harvest, estimation, cultivation practices and climatic factors. Future thrusts of seed oil improvement have been indicated.

QUESTIONS

1. Describe in brief quality of cotton seed oil.

2. Discuss briefly variability found for seed oil content in cotton germplasm.

3. Give a brief comparison of genetic research on cotton seed oil.

4. Explain various methods of estimating seed oil content in cotton.

5. What are the factors affecting seed oil content in cotton?

6. Discuss future thrusts of research on cotton seed oil content.

7. Describe availability of cotton seed for crushing purpose.

8. Explain various sources of high seed oil contents in cotton.

9. Give a brief account of various breeding methods used for improvement of seed oil content in cotton.

Chapter 12

Indian Seed Legislation

INTRODUCTION

The seed can be defined in two ways, *viz.*, (i) in broad sense and (ii) in strict sense. These are explained as follows:

(i) Broad Sense

In the broad sense, any plant part which is used for commercial multiplication of a crop is called seed. For example, in case of sugarcane and potato stem and tuber are also known as seed.

(ii) Strict Sense

In strict sense, seed is the product of fertilized ovule that consists of embryo, seed coat and cotyledon (s).

Improved Seed

The seed of a released and popular variety produced by scientific method is referred to as improved seed or quality seed. Variety refers to a genotype which has been released for commercial cultivation either by State Variety Release Committee or Central Variety Release Committee. Improved seed plays an important role in maximizing production and productivity of field crops. It results in (i) better germination, (ii) vigorous seedling growth, (iii) higher crop stand, (iv) better quality of produce, and (v) ultimately in higher crop yield.

Seed Act

Seed Act refers to the legal procedures approved by the government for production and marketing of seeds. The main objective of the Seeds Act is to ensure availability of quality seeds to farmers. In India, the Seed Act was first introduced in 1966 which is called as Seeds Act 1966. This Act was amended in 1972 which is known as the Seeds amended Act 1972. Subsequently, the seed act was thoroughly revised and the revised act was approved by the Government of India in September 2004. The revised Seeds Act is known as Seeds Act 2004 which came into force with effect from January 2005. It replaces the Seeds Act, 1966.

BRIEF HISTORY

The Indian seeds act has undergone several changes over the years. Significant milestones in the development of Indian seed industry and Indian seed legislation are presented in Table 12.1.

TABLE 12.1: Significant Milestones in the Development of Indian Seed Industry/Legislation

Year	Important Event	Remarks/Brief Description
1963	Establishment of National Seeds Corporation in New Delhi.	NSC was established to take up the work of quality seed production and promote seed industry in the country.
1966	The first Indian Seed Act was formulated	To provide good quality planting seeds to the farmers.
1968	Seed Rules were framed to implement various legislations given Seed Act 1966.	Enactment of Seed Act 1966.
1972	The Seed Act 1966 was amended known as seed amendment Rules 2002.	The Jute seeds were included to the Seeds Act, establishment of Seed Certification Board and fixing of minimum standards.
1973	The Seed Act 1966 was amended known as seed amendment Rules 2003.	Power of appellate authority and duty of seed analyst were slightly modified. The Seed Testing Manual was published by ICAR for reference.
1974	The Seeds [amendment] Rules 1974 were introduced.	This amendment conferred more powers on seed inspectors during crop failure.
1976	The National Seed Project was launched by ICAR, New Delhi.	The NSP was launched to establish State Seeds Corporations in different states.
1978	The UPOV Seed Act was introduced	The new variety should have distinctiveness, uniformity and stability for protection under this act.
1981	The Seeds [amendment] Rules 1981 were introduced	Adoption of Minimum Indian Seed Certification Standards published by Central Seed Committee was made compulsory for seed certification.
1983	The Seeds [control] Order 1983 was introduced.	This Order included Seeds as essential commodity item under the Commodity Act 1955.
1988	The New Policy on Seed Development was introduced	The new policy was formulated to provide Indian Farmers with access to the best available seeds and planting materials of domestic as well as exotic origin.
1989	Plants, Fruits and Seeds Order 1989 was introduced	This replaces the Plants, Fruits and Seeds Order 1984 and provides regulations for post entry quarantine checks.

Year	Important Event	Remarks/Brief Description
1991	The UPOV Seed Act 1991 was introduced	The new variety should have novelty in addition to distinctiveness, uniformity and stability for protection under this modified UPOV Act.
2001	The Protection of Plant varieties and Farmers Rights Act was formulated.	A new variety with novelty, distinctiveness, uniformity and stability can be registered for protection under this Act.
2002	The National Seed Policy was formulated.	The National Seed Policy was formulated to raise Indias' share in the global seed trade by facilitating advanced scientific aspects such as biotechnology to farmers and in March 2002 first transgenic cotton was approved for commercial cultivation in India.
2003	Plant Quarantine [Regulation of Import into India] was introduced.	This Order has now replaced the Plants, Fruits and Seeds Order 1989.
2003	Protection of Plant Varieties Rules 2003 were introduced.	These rules were introduced for smooth implementation of the PPV and FR Act 2001.
2004	The New Seed Bill was formulated and introduced.	This Seed Bill has replaced the Seeds Act 1966.Registration of all seed materials is compulsory and self-certification is permitted.
2005	Seed Act 2004 come into force.	Seed Act 2004 replaced Seed Act 1966.

MAIN ISSUES

The main issues of Indian Seed Act 2004 in relation to coverage, seed category, license, registration, criteria for protection, period of protection, certification, compensation, farmers' privilege, penalty, exclusion, research exemption and constitution of central seed committee are briefly presented below:

1. Coverage

This Act has wider coverage than seeds act of 1966. It includes agriculture, horticulture, forestry, plantation crops, medicinal and aromatic plants. The seeds of all these crops are covered by the new act.

2. Seed Category

This act permits sale of certified seed only. The name of variety, physical purity and germination percentage have to be indicated on the seed container or bag.

3. License

The seed dealers, sellers and growers should have license which is compulsory.

4. Registration

Registration for all kinds and varieties of seed is compulsory. No producer can grow seeds unless he is registered. It is compulsory for all cultivated varieties of field crops, vegetable crops, fruit crops, medicinal and aromatic plants, forest species and plantation crops. Every seed producer and dealer, and horticulture nursery has to be registered with the state government.

5. Criteria for Registration

The released and notified varieties can be registered for seed multiplication. There are four criteria for protection of varieties under PPV and FR Act. These are novelty, distinctiveness, uniformity and stability.

6. Period of Registration

The Indian Seeds Act 2004 has adopted UPOV Act 1978. According to this Act, the period of protection is 15 years for annual and biennial crops and 18 years for perennial plants such as trees and vines.

7. Certification

The Act permits self certification of seeds by accredited agencies and allows the central government to recognize certification by foreign seed certification agencies. Seed producers are permitted to self certify the performance of their seed under certain conditions. The seed companies need to provide the results of multiplication trials before registration.

8. Compensation

The disputes about the quality of seeds have to be settled in the consumer court. Any loss in the production due to poor seed quality is claimed in the consumer court.

9. Farmers' Privilege

The Seeds Act 2004 allows farmers' privilege. The Bill does not restrict the farmer's right to use or sell his farm seeds and planting material, provided he does not sell them under a brand name. All seeds and planting material sold by farmers will have to conform to the minimum standards of germination, physical purity and genetic purity applicable to registered seeds.

10. Penalty

Any person who contravenes any provisions of the Act or imports, sells or stocks seeds deemed to be misbranded or not registered, can be punishable by a fine between Rs. 5,000 and Rs. 25,000. The penalty for giving false information is a prison term up to six months and/or a fine up to Rs. 50,000. In case of companies, the person(s) in charge of conduct of the business of the company will be held accountable.

11. Exclusion

Those techniques which are harmful to animals and the environment such as terminator technology and traitor technology are excluded and not allowed by the seed act.

12. Research Exemption

This act allows use of protected material by breeders for development of new varieties.

13. Central Seed Committee

In the Seed Act 2004, the central seed committee consists of Chairman [Secretary DARE], Agricultural Commissioner, DDG horticulture [ICAR], Joint Secretary Seeds[DAC], representatives from DBT, Ministry of Environment and Forest, Secretary Agriculture [from five States], Director [SSCA] from one State and MD-SSC from one State. In addition, two representatives each of farmers and seed industry are nominated.

CHALLENGES

1. Reduction in Biodiversity

The new seeds act will encourage cultivation of new varieties. Thus old varieties and land races will be replaced by new varieties on vast areas resulting in significant reduction in biodiversity. Reduction in biodiversity will invite danger of uniformity and lead to narrow genetic base and narrow adaptation.

2. Monopoly of Seed Companies

It will encourage multinational seed companies which may lead to monopoly of such companies. Seed companies may hike seed prices resulting in exploitation of farmers.

3. Difficult to Trace Defective Seed Lot

The Seed Act does not provide for a mechanism to trace back a packet of seed to the dealer, processor and producer. This would make it difficult to trace back a defective lot, and rectify any deficiencies in the supply chain.

4. False Certification

Seed producers would be permitted to self-certify the performance of their seeds under certain conditions. This opens up the possibility of false declaration by seed companies. To prevent this, only government agencies should be allowed to conduct these trials and grant certification.

5. Maintenance of Records

Every horticultural nursery has to be registered with the state government and has to maintain records of layout plan, source of every planting material, *etc*. Such nurseries in the unorganized sector may find it difficult to adhere to these conditions.

COMPARISON OF SEEDS ACT 1966 AND SEEDS ACT 2004

There are some similarities and some dis-similarities between two seeds acts. A brief comparison of seed act 1966 and seed act 2004 is presented in Table 12.2.

TABLE 12.2: Comparison of Seed Act 1966 and Seed Act 2004

Sl.No.	Particulars	Seeds Act 1966	Seeds Act 2004
A	**Similarities**		
1	Category of seed permitted for sale	Certified Seed	Certified Seed
2	License	Compulsory	Compulsory
3	Certification	Compulsory	Compulsory
4	Farmers' exemptions	Available	Available
5	Purity and germination standards	Adopted	Adopted
6	Compensation of loss by use of certified seed	Through consumer court	Through consumer court
7.	Requirements for seed multiplication	Released and notified varieties	All seeds for sale must be registered
B	**Dissimilarities**		
1.	Crop covered	Agriculture and horticulture	Agriculture, horticulture, forestry, plantation crops, medicinal and aromatic plants
2	Registration of transgenic varieties	No provision	Special provision
3	Registration with PPVFR authority	Not required	Required
4	Period of protection	Not defined	Defined
5	Penalty for violation of Act	Rs 100-1000/and 6 months imprisonment	Rs 500-50,000/and six months imprisonment
6	Self Certification	Not permitted	Permitted
7	Seed multiplication of land Races	Permitted	Not permitted
8	Representatives in central seed committee	From all States	From five States only
9	Involvement of Private Sector	No	Yes

COMPARISON OF SEEDS ACT 2004 AND PPV & FR ACT 2001

There are some similarities and some dissimilarities between Seed Act 2004 and Protection of Plant Variety and Farmers' Rights Act 2001 which are presented in Table 12.3.

TABLE 12.3: Comparison of Seeds Act 2004 and Protection of Plant Varieties and Farmers' Rights Act 2001

Sl.No.	Particulars	Seeds Act 2004	PPV and FR Act 2001
A	**Similarities**		
1	Registration	Compulsory	Compulsory
2	Period of registration	15 years for annual and biennial crops and 18 years for perennials	15 years for annual and biennial crops, 18 years for perennials

Sl.No.	Particulars	Seeds Act 2004	PPV and FR Act 2001
3.	Renewal of registration	Permitted for similar period	Permitted for similar period
4	Farmers Rights	Allowed	Allowed
5	Plant breeders' Rights	Available	Available
6	Disclosure of variety's performance	Compulsory	Compulsory
7	Sale of branded seeds by farmers	Not permitted	Not permitted
B	**Dissimilarities**		
1	Record Maintained	National Register of Seeds	National register of plant varieties
2	Compensation	Through consumer court	Through PPV and FR Authority
3	Disclosure of variety's parentage	Not required	Required
4	Variety's performance to be declared by	Seed producer, distributor or vendor	Breeder of the variety
5	Penalty for violation of the Act	Rs. 5000-50,000	Rs. 50,000-20 lakh
6	Imprisonment for violation of Act	Up to six months	From 2-3 years

MERITS OF SEED ACT 2004

1. **Wider Coverage**: This act has wider coverage [agriculture, horticulture, forestry, plantation crops, medicinal and aromatic plants] than Seed Act of 1966.

2. **Easy Certification:** This act has made the process of certification very simple means allowed self certification. But it should not be misused.

3. **Bans Harmful Technologies**: The act did not permit registration of harmful technologies such as terminator technology and traitor technology.

4. **Permits Farmers' Rights:** It permits farmers' rights. The farmers need not to register their varieties.

5. **Ensures Quality Seed:** This act ensures availability of good quality seeds to farmers by imposing heavy penalties on the sale of spurious seeds.

6. **Registration for long Duration**: This act provides registration for 15 years for annual and biennial crops and 18 years for perennial crops which is sufficiently long period. Moreover, the registration is renewable for another term of similar duration.

7. **Central Seed Committee:** In the Seed Act 2004, the constitution of central seed committee is better than that of Seed Act 1966.

8. **Promotes Seed Industry:** The new seed act will lead to fast development of seed industry due to involvement of representatives of multinational seed companies or industries in the central seed committee.

9. **Increase in Productivity:** The new act will help in increasing crop productivity due to availability of better quality seeds.

10. The present seed replacement rate is 20-25 per cent. The seed act will enhance seed replacement rate of various crops.

11. The seed act will boost the export of seeds and encourage import of useful germplasm for use in crop improvement programs.

12. It will encourage investment in seed research and development.

DEMERITS OF SEED ACT 2004

1. Although farmers are exempt from registering their seed varieties, the seeds have to conform to standards prescribed for commercial seeds. Farmers may find it difficult to adhere to the standards required of commercially sold seeds, because the seed saved and exchanged by farmers constitute above 80 per cent of the seeds planted

2. It is not clear whether the compensation would include the value of the crop or only the cost of the seed.

3. It is not clear whether a seed producer may sell seed which is registered by a different producer.

4. The disclosure of parentage is not essential. It may lead to a situation where seeds can be registered without disclosing the parentage or origin of the seed.

5. The Act does not have the provision of benefit sharing unlike the Convention on Biological Diversity and the PPV and FR Act.

6. The Seed Inspector has the power to enter and search as well as break open container or break open doors, without a warrant. This is against the normal procedure.

SUMMARY

Seed Act refers to the legal procedures approved by the government for production and marketing of seeds. The main objective of the Seeds Act is to ensure availability of quality seeds to farmers. In India, the Seed Act was first introduced in 1966 which is called as Seeds Act 1966. This Act was amended in 1972 which is known as the Seeds amended Act 1972. Subsequently, the seed act was thoroughly revised and the revised act was approved by the Government of India in September 2004. The revised Seeds Act is known as Seeds Act 2004 which came into force with effect from January 2005. It replaces the Seeds Act, 1966. The significant milestones in the history of Indian Seed Legislation have been highlighted.

The new seed Act of 2004 has wider coverage than seeds act of 1966. It includes agriculture, horticulture, forestry, plantation crops, medicinal and aromatic plants. Main features, merits and demerits of Seed Act 2004 are presented. A comparison of Seeds Act 2004 with Seeds Act 1966 and PPV and FR Act 2001 is presented in tabular form. The future challenges have been outlined.

QUESTIONS

1. Define Seed Law and describe in brief history of seed legislation in India.

2. Discuss briefly main features of Indian Seed Act 2004.

3. Give a brief comparison of Indian Seed Act 1966 and Seed Act 2004.

4. Compare Indian Seed Act 2004 and PPV and FR Act 2001.

5. What are advantages of Indian Seed Act 2004?

6. What are demerits of Indian Seed Act 2004?

7. Discuss briefly challenges related to Indian Seed Act 2004.

8. Write short notes on the following:
   (i)   Terminator technology   (ii) Traitor technology
   (iii) Self-certification   (iv) Improved seed

9. What are differences between Seed Act 1966 and Seed Act 2004?

10. What are differences between Seed Act 2004 and PPV and FR Act 2001?

Appendices

APPENDIX 1: List of Cotton Varieties and Hybrids Released in India

Punjab State

Sl.No.	Varieties/Hybrids	Species	Year of Release	Yield (q/ha)	Duration (days)	GOT (per cent)	MHL (mm)	Spinning Counts	Area of Cultivation
	Varieties								
1.	LH 900	H	1985	27	170	34	22	30	Punjab
2.	LH 886	H	1988	26	165	35	22	30	Punjab
3.	LH 1556	H	1996	21	170	34	27	40	Punjab
4.	F 846	H	1993	26	170	35	23	30	Punjab
5.	F 1378	H	1997	30	175	35	23	30	Punjab
6.	F 505	H	1986	24	170	34	22	30	Punjab
7.	F 1054	H	1993	26	170	36	23	30	Punjab
8.	LH 1134	H	1990	22	170	35	27	40	Punjab
9.	LD 327	A	1987	20	170	41	16	8	Punjab
10.	LD 491	A	1996	14	175	39	20	10	Punjab
11.	LD 694	A	2001	25	175	41	18	10	Punjab
	Hybrids								
12.	Fateh	HH	1994	30	180	34	25	30	Punjab
13	LDH 11	AA	1994	20	175	36	22	20	Punjab
14.	LHH 144	HH	1998	28	180	35	28	50	Punjab
15.	PAU 626	AA	2007	21	165	40	20	NS	Punjab

Haryana State

Sl. No.	Varieties/Hybrids	Species	Year of Release	Yield (q/ha)	Duration (days)	GOT (per cent)	MHL (mm)	Spinning Counts	Area of Cultivation
	Varieties								
1.	HS 6	H	1991	22	175	36	22	30	Haryana
2.	H 974	H	1991	21	170	35	23	30	Haryana
3.	H 1098	H	1995	19	175	35	25	30	Haryana
4.	HS 182	H	1997	22	165	36	21	30	Haryana
5.	H 1117	H	2001	22	180	34	23	30	Haryana
6.	HD 107	A	1996	26	175	38	18	10	Haryana
7.	HD 123	A	1997	23	165	39	18	10	Haryana
8.	HD 324	A	2005	22	175	42	16	NS	Haryana
9.	CISA 310	A	2007	21	170	37	20	NS	Haryana
10.	H 1226	H	2007	24	165	35	25	30	Haryana
	Hybrids								
11.	Dhanlaxmi	HH	1994	25	180	35	26	40	Haryana
12.	Om Shankar	HH	1996	28	165	34	25	40	North zone
13.	AAH 1	AA	1999	24	180	38	16	> 10	Haryana
14.	CISAA 2	AA	2004	25	175	38	20	10	North zone
15.	CSHH 198	HH	2004	27	165	35	24	40	North zone
16.	HHH 287MS	HH	2005	24	165	34	27	40	Haryana
17.	CSHH 238	HH	2007	22	165	33	28	40	North Zone

Rajasthan State

Sl. No.	Varieties/Hybrids	Species	Year of Release	Yield (q/ha)	Duration (days)	GOT (per cent)	MHL (mm)	Spinning Counts	Area of Cultivation
	Varieties								
1.	RST 9	H	1991	26	175	36	23	30	North Rajasthan
2.	RS 875	H	1997	30	175	35	23	30	North Rajasthan
3.	G.Ageti	H	1978	14	190	33	23	28	Sriganganagar
4.	RS 810	H	2000	20	170	35	23	30	North Zone
5.	RG 8	A	1986	17	180	39	16	9	Sriganganagar area
6.	RG 18	A	2000	22	170	38	18	15	North Zone
	Hybrids								
7.	Maruvikas	HH	1994	30	180	34	24	30	Rajasthan
8.	Raj DH 7	AA	2001	28	170	39	23	20	Rajasthan

Madhya Pradesh

Sl. No.	Varieties/Hybrids	Species	Year of Release	Yield (q/ha)	Duration (days)	GOT (per cent)	MHL (mm)	Spinning Counts	Area of Cultivation
	Varieties								
1.	Khandwa-2	H	1971	8R	160	36	24	30	Nimar region
2.	Khandwa-3	H	1984	9R	180	34	23	36	Nimar region
3.	Vikram	H	1981	9R	160	33	24	35	Malwa region
4.	JK 4	H	1999	18	160	34	26	35	Nimar region
5.	Maljari	A	1954	6R	180	33	21	20	Malwa and Nimar tracts
6.	Jawahar Tapti	A	1992	15	150	35	24	30	East and West Nimar
7.	Sarvottam	A	2001	20	15	34	24	25	Nimar
	Hybrids								
8.	JKHY 1	HH	1976	25	210	35	27	50	M.P. and A.P.
9.	JKHY 11	HB	1982	18	240	31	31	60	Irrigated areas of M.P.
10.	JKHY 2	HH	1997	30,15R	180	34	27	50	M.P.

Maharashtra State

Sl. No.	Varieties/Hybrids	Species	Year of Release	Yield (q/ha)	Duration (days)	GOT (per cent)	MHL (mm)	Spinning Counts	Area of Cultivation
	Varieties								
1.	DHY 286	H	1975	10R	200	36	25	40	Vidarbha region
2.	Rajat	H	1994	12R	175	39	25	40	Vidarbha region
3.	LRA 5166	H	1982	10R	180	35	24	34	South zone and MS
4.	LRK 516	H	1992	12R	160	36	25	36	MS, Gujarat
5.	AKH 4	A	1975	7R	190	39	24	30	Vidarbha and Marathwada
6.	AKA 5	A	1981	7R	180	39	22	30	Vidarbha and Marathwada
7.	AKA 8401	A	1989	10R	200	38	25	40	Vidarbha region
8.	PA 183	A	1996	18	180	39	27	35	Marathwada region
9.	Y1	A	1962	5R	190	39	24	30	Khandesh
10.	AKA 7	A	1998	10R	150	41	22	25	Vidarbha region
11.	PA 402	A	2003	18R	165	37	27	30	Maharashtra
12.	AKA 8	A	2005	12R	190	38	26	35	Vidarbha
13.	AKH 8828	H	2005	12R	180	42	27	40	Vidarbha

Sl. No.	Varieties/Hybrids	Species	Year of Release	Yield (q/ha)	Duration (days)	GOT (per cent)	MHL (mm)	Spinning Counts	Area of Cultivation
	Hybrids								
14.	Godavari (NHH 1)	HH	1978	15 R	180	35	28	50	Marathwada region
15.	Savitri (RHR 253)	HB	1978	28	175	32	30	60	Deccan Canal Area
16.	PKV Hy2	HH	1981	12 R	180	36	27	40	M.S,Vidarbha
17.	NHH 44	HH	1983	23	180	35	24	50	Marathwada
18.	RHH 195 (Sampada)	HH	1986	21	160	36	24	40	Deccan Canal Area
19.	NHB 12	HB	1989	30	180	33	33	80	Marathwada
20.	CICR HH1	HH	1991	25	185	35	25	36	Marathwada
21.	NHH 302	HH	1991	20	170	35	25	40	Marathwada
22.	PKV Hy3	HH	1994	15 R	180	37	25	40	Marathwada and Gujarat
23.	PKV Hy4	HH	1996	20	165	35	28	50	Vidarbha region
24.	AKDH 7	AA	2001	15	170	36	22	20	Vidarbha
25.	PKV Hy5	HH	1999	15	170	36	25	40	Vidarbha
26.	Pha 46	HA	1996	17	180	32	28	40	Marathwada
27.	AKDH 5	AA	2006	12R	180	38	25	35	Vidarbha

Gujarat State

Sl. No.	Varieties/Hybrids	Species	Year of Release	Yield (q/ha)	Duration (days)	GOT (per cent)	MHL (mm)	Spinning Counts	Area of Cultivation
	Varieties								
1.	G.Cot 12	H	1981	6R	220	36	24	23	Wagad area
2.	G.Cot 16	H	1995	16R	140	37	25	40	Middle Gujarat
3.	G.Cot 18	H	2000	30	160	36	25	30	Junagdh,Saurashtra
4.	G.Cot 15	A	1989	16R	150	38	22	30	Mathio tract
5.	G.Cot 19	A	1997	11R	140	34	23	30	Mathio tract
6.	G.Cot 13	h	1981	8R	190	39	23	30	Wagad area
7.	G.Cot 17	h	1995	11R	210	37	23	40	Middle Gujarat
8.	G.Cot 21	h	1998	11R	215	42	22	30	Wagad area
9.	G.Cot 23	h	2000	13R	200	39	22	30	Rainfed area
	Hybrids								
10.	H 4	HH	1970	35	230	33	28	50	Gujarat, A.P., K.S., M.S.
11.	H 6	HH	1980	35	210	34	27	60	Gujarat, M.S., A.P.
12.	DH 7	hA	1985	15 R	190	37	22	30	Gujarat
13.	DH 9	hA	1988	15 R	190	34	28	40	Gujarat,M.P.
14.	H 8	HH	1989	35	180	36	28	50	Gujarat
15.	H 10	HH	1995	18 R	150	35	26	40	Gujarat
16.	G.Cot MDH 11	AA	2001	27	140	36	24	20	Gujarat
17.	G. Cot HB 102	HB	2004	15R	200	34	32	80	Gujarat
18.	G. Cot H 12	HH	2006	18R	175	34	28	40	All India

Andhra Pradesh and Telangana

Sl. No.	Varieties/Hybrids	Species	Year of Release	Yield (q/ha)	Duration (days)	GOT (per cent)	MHL (mm)	Spinning Counts	Area of Cultivation
	Varieties								
1.	Suvin	B	1974	15	190	30	36	120	Irrigated Areas
2.	NA 247	H	1982	8R	150	35	25	40	Rayalseema
3.	NA 920	H	1988	25	170	38	24	34	Rayalseema
4.	L 389	H	1993	25	170	35	27	50	NSP Tract
5.	L 603	H	1997	23	155	35	28	40	NSP Tract
6.	L 604	H	1997	26	160	36	26	40	Rice fallow area
7.	Kanchana	H	1987	28	170	34	25	40	Whitefly prone area
8.	LK 861	H	1993	25	165	35	24	40	Whitefly prone area
9.	NA 1325	H	1993	8R	180	36	24	30	Rainfed area
10.	Srisailam	A	1976	5R	150	35	22	26	Kurnool area
11.	Mahanandi	A	1978	5R	180	32	22	26	Rainfed area
12.	Sahana	H	1996	20	170	40	26	40	NSP Area
13.	Surabhi	H	1997	35	170	35	29	55	NSP Area
14.	Sumangala	H	2000	25	165	37	25	40	NSP Area
15.	LRA 5166	H	1982	10R	180	35	24	34	Rainfed Area
16.	Raghvendra	h	1997	12R	190	36	25	25	Rayalseema
17.	Arvinda	A	1997	12R	180	36	26	30	Rayalseema

Sl. No.	Varieties/Hybrids	Species	Year of Release	Yield (q/ha)	Duration (days)	GOT (per cent)	MHL (mm)	Spinning Counts	Area of Cultivation
	Hybrids								
18.	NHB 80	HB	1982	20	170	34	27	50	NSP area of A.P.
19.	LAHH 1	HH	1987	28	150	35	29	60	A.P.
20.	LAHH 4	HH	1997	30	170	35	31	40	All 3 zones
21.	H6	HH	1980	35	210	34	27	60	Whole A.P.
22.	JKHY 1	HH	1980	25	210	35	27	50	Whole A.P.
23.	DHH 11	HH	1996	30	190	35	37	50	Whole A.P.
24.	Savita	HH	1987	30	170	34	30	60	Whole A.P.
25.	Surya	HH	1997	25	170	38	31	60	Whole A.P.
26.	Varalaxmi	HB	1974	30	210	34	31	80	NSP Area
27.	DCH 32	HB	1981	35	190	36	33	80	NSP Area
28.	NHB 80	HB	1982	20	170	34	27	50	NSP Area
29.	DHB 105	HB	1996	30	190	34	30	80	NSP Area
30.	Sruthi (HB)	HB	1997	30	150	33	37	80	NSP Area
31.	DDH 2	Ah	1992	12R	180	34	22	20	Rainfed Areas

Karnataka State

Sl. No.	Varieties/Hybrids	Species	Year of Release	Yield (q/ha)	Duration (days)	GOT (per cent)	MHL (mm)	Spinning Counts	Area of Cultivation
	Varieties								
1.	Sharda	H	1981	12R	180	38	28	40	Rainfed areas
2.	Abadhita	H	1983	25	180	37	27	50	Karnataka
3.	Sahana	H	1996	20	170	40	26	40	South zone
4.	DB 3-12	h	1979	4R	170	33	22	30	North East Karnataka
5.	Raichur 51	h	1963	3R	200	34	21	26	Karnataka
6.	DLSA 17	A	2003	15R	165	35	27	30	Karnataka
7.	RAHS 14	H	2003	07R	210	32	22	20	North Karnataka
8.	RAHS 100	H	2004	20	160	40	27	40	North Karnataka
	Hybrids								
9.	Varalaxmi	HB	1972	30	210	35	31	80	South zone and M.S.
10.	DCH 32	HB	1981	35	190	36	33	80	South zone
11.	DDH 2	hA	1992	12 R	180	34	22	20	South zone
12.	DHB 105	HB	1996	30	190	34	33	80	South zone
13.	DHH 11	HH	1996	30	190	35	27	50	South zone
14.	DDHC 11	AA	2003	7R	190	32	22	20	N. Karnataka
15.	RAHH 95	HH	2006	30	165	40	27	50	N. Karnataka
16.	RAHB 87	HB	2006	25	200	36	35	80	N. Karnataka

Tamil Nadu

Sl. No.	Varieties/Hybrids	Species	Year of Release	Yield (q/ha)	Duration (days)	GOT (per cent)	MHL (mm)	Spinning Counts	Area of Cultivation
	Varieties								
1.	MCU 7	H	1972	12R	145	35	23	30	Rice fallows
2.	MCU 5 VT	H	1982	20	165	34	29	50	Winter tract
3.	LRA 5166	H	1982	10R	180	35	24	34	South Zone and MS
4.	Surabhi	H	1997	35	170	35	29	55	South zone
5.	Sumangala	H	2000	25	165	37	25	40	South zone
6.	MCU 12	H	2000	30	150	37	27	40	Tamil Nadu
7.	SVPR 2	H	1996	16R	160	36	25	30	Summer tract
8.	Suvin	B	1974	15	190	30	36	120	Tamil Nadu
9.	K 10	A	1984	7R	145	38	24	30	Tamil Nadu
10.	K 11	A	1992	8R	135	35	24	30	Tamil Nadu
11.	KC 3	H	2006	11R	150	36	26	30	Tuticorin
	Hybrids								
12.	CBS 156	HB	1974	30	180	32	33	100	Tamil Nadu
13.	Suguna	HH	1978	30	150	35	25	40	Tamil Nadu
14.	KCH 1	HB	1980	30	150	34	31	60	Tamil Nadu
15.	Savita	HH	1987	30	170	34	30	60	T.N. and A.P.
16.	HB 224	HB	1989	30	170	33	31	80	Tamil Nadu
17.	TCHB 213	HB	1990	30	190	32	33	80	Tamil Nadu
18.	Surya	HH	1997	25	170	38	31	60	South zone
19.	Sruthi	HB	1997	30	150	33	37	80	South zone

APPENDIX 2: Public Sector Cotton Varieties and Hybrids under Seed Production Chain

Varieties

Sl.No.	Name of Variety	Year of Release	Released From	Recommended for
1	AKH-5 (AKH-605)	1983	Akola	Vidarbha
2	AKH-7 (AKA 8307)	2001	Akola	Vidarbha
3	AKA-8	2008	Akola	Vidarbha
4	AKH-081	1988	Akola	Vidarbha
5	AKH-8828	2008	Akola	Vidarbha
6	Anusaya (NH-615)	2008	Guntur	Telangana
7	ARVINDA (NDL-2708)	2000	Guntur	Telangana
8	CICR-1 (CISA-310)	2010	Sirsa	Haryana
9	CICR-3 (CISA-614)	2010	Sirsa	Haryana
10	CNHO 12	2010	CICR, Nagpur	Vidarbha
11	DDHC-11	2008	Dharwad	Karnataka
12	DLSA-17	2009	Dharwad	Karnataka
13	F 1861	2004	Faridkot	Punjab
14	F-1054	1993	Faridkot	Punjab
15	F-1378	1997	Faridkot	Punjab
16	F-505	1987	Faridkot	Punjab
17	F-846	1993	Faridkot	Punjab
18	FDK 124	2011	Faridkot	Punjab
19	G. Cot 19 (G.Am-31)	2000	Surat	Gujarat
20	H 1236	2010	Hisar	Haryana
21	H-1098	1997	Hisar	Haryana
22	H-1117	2002	Hisar	Haryana
23	H-1226	2007	Hisar	Haryana
24	H-1300	2012	Hisar	Haryana
25	HD 432	2010	Hisar	Haryana
26	HD-123	2000	Hisar	Haryana
27	JAWAHAR KAPAS-3	1998	Khandwa	Khandwa Tract
28	JAWAHAR KAPAS-4 (JK-4)	2002	Khandwa	Khandwa Tract
29	JAWAHAR KAPAS-5 (JK-5)	2007	Khandwa	Khandwa Tract
30	KC-3	2007	Khandwa	Khandwa Tract
31	L-603	2000	Lam	P.
32	L-604	2000	Lam	Andhra Pradesh
33	LH-1556	2010	Ludhiana	Punjab
34	LH 2076	2010	Ludhiana	Punjab
35	LRA-5166	1983	Coimbatore	TN, AP, MS, MP

Sl.No.	Name of Variety	Year of Release	Released From	Recommended for
36	LUDHIANA DESI-327	1989	Ludhiana	Punjab
37	MCU-12 (TCH-1025)	2000	TNAU	Tamil Nadu
38	MCU-5 VT	1984	TNAU	Tamil Nadu
39	Nandyal-1	2007	Nandyal	P.
40	NH-452	1996	Nandyal	P.
41	NH-545	2004	Nandyal	P.
42	PA-402	2005	Parbhani	Marathwada
43	GN Cot 22	2013	Surat	Gujarat
44	G Cot 20	2014	Surat	Irrigated GS
45	GN Cot. 25	2010	Bharuch	Rainfed Gujarat
46	Suvin	1974	CICR, Coimbatore	Irrigated Tamil Nadu

Hybrids

Sl.No.	Name of Hybrid	Year of Release	Released From	Area of Cultivation
1	AAH-1 (Desi Cotton Hybrid-1)	1999	CICR, Sirsa	Haryana
2	DHH-543 (Suvidha)	2008	Dharwad	Karnataka
3	HHH-223	2002	Hisar	Haryana
4	Lam Cotton Hybrid-7	2007	Guntur	Andhra Pradesh
5	Moti (LMDH-8)	2007	Ludhiana	Punjab
6	NHH-44	1985	Nanded	Marathwada
7	PHULE-492 (RHH-0492)	2002	Rahuri	Western Maharashtra
8	PKHY-2	1983	Akola	Vidarbha
9	RAHB-87	20098	Rahuri	Western Maharashtra
10	RAHH-95	2009	Rahuri	Western Maharashtra
11	RAHH-98	2009	Rahuri	Western Maharashtra
12	RAJDH-9	2006	Ganganagar	Rajasthan
13	Shresth (CSHH-198)	2005	CICR, Sirsa	Haryana
14	G. Cot Hybrid 6 [BG II]	2012	Surat	GS, MS, MP
15	G. Cot Hybrid 8 [BG II]	2012	Surat	GS, MS, MP
16	G. Cot Hybrid 10 [BG II]	2015	Surat	GS, MS, MP
17	G. Cot Hybrid 12 [BG II]	2015	Surat	GS, MS, MP
18	GN Cot Hybrid 14	2014	Surat	Irrigated GS

APPENDIX 3: (a) Area, Production and Yield of Cotton in India during 2015-16 [CAB]

Sl.No.	State	Area L/ha	Production L/Bales	Lint Yield kg/ha
1	Punjab	3.39	7.50	376
2	Haryana	6.03	15.00	423
3	Rajasthan	4.48	15.00	569
4	North total	13.90	37.50	459
5	Gujarat	27.19	94.00	588
6	Maharashtra	38.27	75.00	333
7	Madhya Pradesh	5.47	18	559
8	Central total	70.93	187.00	448
9	Telangana	17.78	59.50	569
10	Andhra Pradesh	6.66	24.00	613
11	Karnataka	6.33	20.00	537
12	Tamil Nadu	1.42	5	599
13	South Total	32.19	108.50	573
14	Orissa	1.25	3.00	408
15	Others	0.50	2.00	680
	TOTAL	118.77	338.00	484

APPENDIX 3: (b) Area, Production and Yield of Cotton in India during 2016-17 [CAB]

Sl.No.	State	Area L/ha	Production L/Bales	Lint Yield kg/ha
1	Punjab	2.56	9	598
2	Haryana	4.98	20.00	683
3	Rajasthan	4.42	18.00	692
4	North total	11.96	47.00	668
5	Gujarat	24.00	95.00	673
6	Maharashtra	38.06	89.00	398
7	Madhya Pradesh	5.99	21.00	596
8	Central total	68.05	205.00	512
9	Telangana	12.50	48.00	653
10	Andhra Pradesh	4.49	19.00	719
11	Karnataka	4.64	21.00	769
12	Tamil Nadu	1.50	6	680
13	South Total	23.13	94.00	691
14	Orissa	1.36	3.00	375
15	Others	0.50	2.00	680
	TOTAL	105.00	351.00	568

APPENDIX 3: (c) Area, Production and Yield of
Cotton in India in different Years

Year	Area L/ha	Production L/Bales	Lint Yield kg/ha
1947-48	44.24	33.36	132
1950-51	58.82	34.3	99
1960-61	76.1	60.12	134
1970-71	76.05	56.64	127
1980-81	78.23	78	169
1990-91	74.39	117	267
2000-01	85.76	140	278
2001-02	87.3	158	308
2002-03	76.67	136	302
2003-04	76.3	179	399
2004-05	87.86	243	470
2005-06	86.77	241	472
2006-07	91.44	280	521
2007-08	94.14	307	554
2008-09	94.06	290	524
2009-10	103.1	305	503
2010-11	111.42	339	517
2011-12	121.78	367	512
2012-13	119.78	370	525
2013-14	119.6	398	566
2014-15	128.46	386	511
2015-16	118.77	338	484
2016-17	105.00	351.00	568

**APPENDIX 4: Some Promising Germplasm Lines of
Upland and *arboreum* Cottons for Various Characters**

Sl.No.	Character	Promising Germplasm Lines	
		Upland Cotton (G. hirsutum)	Desi Cotton (G. arboreum)
1	Earliness	NHS 1412, PKV 081, SIMA 1, T X ORSC 801-79, ACALA 8-1, TAMCOT SP 21, ACALA 69/5, U. ARK, D 203-5, D 238-13-5	AK 14, K 41, AC 13, AC 18, AC 36, AC 48, AC 60, AC 63, AC 65, AC 543, A 733, C 410, DC 92
2	Big boll size	T 120-76, GP 188, DS 56, DS 59, TAMCOT SP 215, EC 174 EBR, 108 F, WIR 3817, NC SMOOTH 1, TASHKENT 1, 133 F, PAKISTAN 2	EC 174092, 30838, 30856, G 135-49, GARO HILL (RF), PM 2-2000, PM 9-2000
3	High ginning outturn	NC 177-16-30, ARKAT 2-1, GP POOL, M 100 SIND, IRMA 23, I 772, STAM 72, NH 572, KIRGHIS K2, HALF and HALF, SENEGAL, VC 45, K 3499	GARO HILL (YF), GARO HILL (RF), 30814, 30838
4	High fibre length	AR 27, DC 534-3, EL 500, EL 592, EL 258, EMPIRE WR, EWLS X TW, G 21-17-619-3, GALAMA, 70 IH 452	LS 1, LS 2, LS 3, ADONICUM, K7, K8, K9, K10
5	High fibre strength	AP 32, JR 48, HANCOCK, DP 16, TH 11, LANKART 57, TEXAS 61, EL 1143 E, END 63, REX, EL 313 W, ACALA 1517 BR	AC 616, AC 630, H 474, GAO 16C B8, H 511, DESI 72, INDICUM 2, INDICUM 38, K 1616-1, OBTUSIFOLIUM INDICA, PBN 48, PBS 6652, DESI 72
6	Jassid resistant	BADNAWAR 1, DELTAPINE 14, MCU 1, BURI NL NAKED, M 100 SIND, LSS 1-2	AC 27, 30797, 30805, DESI 6, MILLION DOLLAR, BDN 6377, ROZI 6, MALVENSIS CHINESE SPOTLESS, CHANDROLLA
7	Whitefly resistant	KANCHANA, LK 861, SUPRIYA, S 505, T X ORHU 1-78, T X ORSC 78, T X MAROON 2-78	OBTUSIFOLIUM COCANADA NLL SP 1, N 723 WR
8	Bollworm resistant/ tolerant	CULTURE 21, DELCERRO, KIANGPO, M 100 SIND, BUSSOGA, CPD 8-1, L 11-A, LO 313 W, EL 174 (YC), M4-58, KAMPALA	AC 27, AC 36, 30811, 30805, AC 28, LD 132, LD 135, G 27, JL 10 BLL, 79/BH-53, B-15-3-MLL
9	Drought resistant	A 59, A 65 F, ARABHAVI, LRA 5166, ARKANSAS 61-28, AP 18-1 (1254), B3-955, EG 3	MOST OF THE DESI GERMPLASM LINES ARE TOLERANT TO DROUGHT
10	Salinity tolerant	PKV 081, KHANDWA 2, KHANDWA 3, JK 345, BADNAWAR, LH 1556, CNH 36, ACALA 44	DESI COTTONS ARE GENERALLY TOLERANT TO SOME DEGREE OF SALINITY

Sl.No.	Character	Promising Germplasm Lines	
		Upland Cotton *(G. hirsutum)*	*Desi Cotton* *(G. arboreum)*
11	Disease resistant	Bacterial blight 101-102 B, REBA B 50, BJA 592, C 2-3 (Y), L-11, LINE 3, C4 B5-8 Alternaria CP 15/2, CPD 8-1, DIXIEKING B, L 11	Fusarium wilt FWR 1, FWR 2, FWR 3, FWR 4, FWR 5 Grey mildew G 135-46, 30805, 30814, 30826, 30838, 30856, EC 1704092 (BANGLADESH)

APPENDIX 5: Zone-wise *Bt.* Cotton Hybrids Approved for Cultivation in India [2009]

(i) North Zone [164 Hybrids, 5 events and 26 Companies]

Bollgard I Hybrids		Bollgard II Hybrids		Other Hybrids
ABCH-3083 Bt.	NCS-901Bt.	ABCH-1299Bt. BGII	RCH-605BG II	Navkar-5Bt.
ABCH-3483 Bt.	NCS-902Bt.	ABCH-2099Bt. BGII	RCH-314BG II	NCEH-6R
ABCH-1857 Bt.	NCS-903Bt.		RCH-134	NCEH-26Bt.
ABCH-172 Bt.	NCS-904Bt.	ABCH-4899Bt. BGII	PRCH-302	NCEH-31Bt.
ABCH-173 Bt.	NCS-905Bt.		PRCH-333	NCH-1005 Bt.
ABCH-174 Bt.	Ole	ABCH-7399Bt. BGII	SDS-26BGII	NCH-1085 Bt.
ABCH-177 Bt.	PCH-401Bt.		SDS-6003BGII	NCH-1163 *Bt.* CH-1177 Bt.
ABCH-178 Bt.	PCH-402Bt.	ABCH-143Bt.BGII	SDS-234BGII	
Ankur 3028 Bt.	PCH-403Bt.	ABCH-146Bt. BGII	SDS-9	UPLHH-271Bt.
Ankur 8120 Bt.	PCH-406Bt.	ABCH-181Bt. BGII	SDS-36	UPLHH-342Bt.
Ankur 651 Bt.	RCH-134	ABCH-182Bt.BGII	SOLAR-56BGII	UPLHH-350Bt.
Ankur 2226 Bt.	RCH-338	ABCH-191Bt. BGII	SOLAR-64BGII	UPLHH-1
Ankur 2534 Bt.	RCH-314	ABCH-192Bt.BGII	SOLAR-65BGII	ZCH-193Bt.
GK 206	RCH-317	ACH-33-2	SOLAR-72BGII	JKCH-1950Bt.
IT 905	SDA-9	Ankur-3028BGII	SOLAR-75BGII	JKCH-99 Bt.
Jai Bt.	SDS-1368	Ankur-5642	SOLAR-76BGII	JKCH-1145Bt.
KDCHH-507 BGI	Shakti-9Bt.	Ankur-8120	SOLAR-77BGII	JKCH-1923Bt.
KDCH-9810	Sigma	GK-212	SO7H-878BGII	JKCH-1945 Bt.
MRC-6025	SP 7007	JaiBGII	SP-1169B2	JKCH-1947
MRC-6029	B-1	Jassi	SP7010B2	JK-1050
MRC-6301	VBCH-1006BG	KCH-36BHII	SWCH-4707BGII	JKCH-226Bt.
MRC-6304	VBCH-1008BG	KCH-999BGII	SWCH-4711BGII	BN Bt[Variety]
NAMCOT-402	VBCH-11BG	KCH-1459BGII	SWCH-2BGII	
NCS-138	6317Bt.	KCH-1539BGII	SWCH-4704BGII	
NCS-913	6488Bt.	KCH-100BHII	SWCH-4713BGII	
NCS-950		KCH-172BGII	Tulsi-162BGII	
		KCH-189BGII	Tulsi-225BGII	
		KCH-311BGII	Tulsi-4BGII	
		KCH-707Bt.	Tulsi-45BGII	
		KDCHH-541BGII	VBCH1515BGII	
		KDCH-441	VBCH-1516BGII	
		MRC-7361BHII	VBCH-1517BGII	
		MRC-7365BGII	VBCH-1518BGII	
		MRC-7017	VBCH1515BGII	

Bollgard I Hybrids	Bollgard II Hybrids		Other Hybrids
	MRC-7031	VBCH-1501	
	MRC-7041	VBCH-1504	
	MRC-7045	VICH-307BGII	
	NAMCOT-616BGII	VICH-308BGII	
	NAMCOT-617BGII	VICH-309BGII	
	NCS-855Bt.2	VICH-310BGII	
	NCS-856Bt.2	VICH-9	
	NCS-857Bt.2	VICH-11	
	NCS-858Bt.2	569	
	NCS-145 (Bunny)	6488-2	
	PCH-876Bt.2	2510-2	
	PCH-877Bt.2	3113-2.	
	PCH-878Bt.2		
	PCH-879Bt.2		
	RCH-602BG II		

(ii) Central Zone [296 Hybrids, 6 events and 35 Companies]

Bollgard I Hybrids		Bollgard II Hybrids		Other Hybrids
ABCH-3083 Bt.	NSPL-405	ABCH-1299Bt. BGII	NSPL-333BGII	ACH-1050Bt.
ABCH-3483 Bt.	NSPL-999		NSPL-432BGII	ACH-1151Bt.
ABCH-1857 Bt.	PCH-404Bt.	ABCH-2099Bt. BGII	NSPL-666BGII	ACH-1171Bt.
ABCH-172 Bt.	PCH-405Bt.	ABCH-4899Bt. BGII	NSPL-36BGII	ACH-1019
ABCH-173 Bt.	PCH-407Bt.		NSPL-405BGII	Dhruv Bt.
ABCH-174 Bt.	PCH-408Bt.	ABCH-7399Bt.	NSPL-999BGII	Kashinath
ABCH-177 Bt.	PCH-409Bt.	BGII ABCH-1020Bt.	Paras Lakshmi	GBCH-07Bt.
ABCH-178 Bt.	PCH-115		PCH-115Bt.2	GBCH-O9Bt.
ABCH-1165	PCH-207	ABCH-143Bt.BGII	PCH-881Bt.2	GBCH-01
ABCH-1220	PCH-923	ABCH-146Bt. BGII	PCH-882Bt.2	Mansoon Bt.
ACH-33-1	PCH-930	ABCH-181Bt. BGII	PCH2171Bt.2	Navkar-5
ACH-155-1	PRCHB-405BG I	ABCH-182Bt.BGII	PCH-205Bt.2	NCEH-2R
ACH-177-1	PRCH-102	ABCH-191Bt. BGII	PRCH-331BtII	NCEH-3R
Akka	PRCH-31	ABCH-192Bt.BGII	PRCH-333Bt2	NCEH-21
Ankur-3042Bt.	Rudra	ACH-111-2	PRCH-504	NCEH-23
Ankur-9	RCH-2	ACH-177-2	PRCH-505	NCEH-14
Ankur-651	RCH-118	Ajeet-11-2	RCH-608BGII	NCEH-34Bt.
Ankur-3032Bt.	RCH-138	Ajeet-155-2	RCH-377BGII	SBCH-286Bt.
		Akka		

Bollgard I Hybrids		Bollgard II Hybrids		Other Hybrids
Ankur HxB-1950Bt.	RCH-144	Amar-1065Bt.	RCH-530BGII	TPHCNO7-015Bt
Brahma	RCH-377	Ankur-3028BG II	RCH-2	TPHCNO7-005Bt
Dyna	RCH-386	Ankur-3034BGII	RCH-515	TPHCNO7-009Bt
GK-204	RCH-395Bt.	Ankur-216BGII	RCH-578	UPLHH-271Bt
GK-205	Sarju BG	Ankur-257BGII	RCH-584	UPLHH-17Bt
Jai Bt.	Sigma	Ankur-3070BGII	Sarju BGII	UPLHH-12Bt
KCH-135	SP-1136	Ankur-HB-2104BGII	SOLAR-66BGII	UPLHH-189Bt
KCH-707	B 1	Atal	SOLAR-60BGII	UPLHH-352Bt
KDCHB-407BG I	SP-499	BrahmaBGII	SP-904B2	UPLHH-13Bt
KDCHH-507BG I	SP-503	GK-218BGII	SP-1016B2	UPLHH-1Bt
KDCHH-786	SP-5-4(Dhanno)	GK-221BGII	SP-1170B2	UPLHH-10Bt
KDCHH-9632	SP-904	GK-224BGII	SP-504	UPLHH-2Bt
KDCHH-9810	SP-923	GK-231BGII	Super-5BGII	YRCH-4Bt.
KDCHH-9821	SWCH-4428Bt.	GK-235BGII	Sudarshan BGII	YRCH-9Bt.
MahasangramBG	SWCH-4531Bt.	GK-205	SWCH-2BGII	YRCH-13Bt.
MECH-12	SWCH-4314Bt.	JaiBGII	SWCH-4708BGII	YRCH-31t.
MECH-162	Tulsi-4	KCH-14K59BGII	SWCH-4715GII	YRCH-45Bt.
MECH-184	Tulsi-5Bt.	KCH-15K39BGII	SWCH-1BGII	YRCH-54Bt.
MRC-6301	Tulsi-9Bt.	KCH-36BGII	SWCH-5017	ZCH-50005
NCS-906Bt.	Tulsi-117	KCH-999BGII	SWCH-5011	ZCH50072Bt.
NCS-907Bt.	VBCH-101	KCH-707	Tulasi-135BGII	JK ChamundiBt.
NCS-908Bt.	VBCH-1006	KCH-135	Tulasi-144BGII	JK Gauri Bt.
NCS-909Bt.	VBCH-1009	KDCHH-541BGII	Tulasi-162BGII	JKCH 2245Bt.
NCS-910Bt.	VBCH-1010	KDCHB-407BGII	Tulasi-117BGII	JKCHB 229Bt.
NCS-138	VBCH-1016	KDCHH-441	Tulasi-4	JK Ishwar
NCS-145(Bunny)	VBCH-1017	KDCHH-621	Tulasi-9	JKCH-99
NCS-207(Mallika)	VCH-111	KDCHH-9632	Tulasi-118	JKCH-226
NCS-913	VICH-5	Krishna-BGII	VBCH1511	JKCH-666
NCS-929	VICH-9	MLBCH-6BGII	VBCH-1516	JK Durga
NCS-950	VICH-15	MLBCH-317	VBCH-1519	JK IndiraBt.
NCS-954	322Bt.	MRC-7373BGII	VBCH-1520	JK Varuna
NCS-955	110Bt.	MRC-7383BGII	VBCH1521	PCH-99Bt.
NCHB-991	6188Bt.	MRC-7301	VBCHB-1525	PCH-77Bt.
NCHB-992	563Bt	MRC-7326	VBCHB-1526	PRCH-712Bt.
NPH-2171	311Bt.	MRC-7347	VICH-311BGII	PRCH-713Bt.
NSPL-36		MRC-7351	VBCH-1501	PRCH-714Bt.

Bollgard I Hybrids		Bollgard II Hybrids		Other Hybrids
		MRC-7918	VBCH-1503	PRCH-715Bt.
		NAMCOT-614BGII	VBCH-1505	MH-5125Bt.
		NAMCOT-615BGII	VICH-312BGII	MH-5174Bt.
		NAMCOT-603BGII	VICH-313BGII	BN Bt.(Variety)
		NAMCOT-605BGII	VICH-314BGII	
		NCS-859Bt.2	VICH-5Bt.	
		NCS-860Bt.2	VICH-15	
		NCS-861Bt.2	311-2	
		NCS-862Bt2	557-2	
		NCS-853Bt.2	110-2	
		NCS-145Bt.2	111-2	
		NCS-207Bt2	195-2	
		NCS-854Bt.2		
		NCHB-945Bt.		

(iii) South Zone [294 Hybrids, 6 events and 35 Companies]

Bollgard I Hybrids		Bollgard II Hybrids		Other Hybrids
ABCH-172Bt.	NCHB-945Bt.	Ankur-3028BGII	PCH-2270	Dhruv Bt.
ABCH-173Bt.	NCHB-990	Ankur-3034BGII	PCH-105	GBCH-04 Bt.
ABCH-174Bt.	NCHB-992	Ankur-257BGII	PRCH-331BG II	GBCH-07Bt.
ABCH-177Bt.	NPH-2171	Ankur-356BGII	PRCH-333BGII	Kashinath
ABCH-178Bt.	NSPL-9	Ankur-3066BGII	PRCH-504	Mansoon Bt.
ABCH-3083Bt.	NSPL-36	AnkurHB-2110 BGII	PRCH-505	NCEH-2R
ABCH-3483Bt.	NSPL-603		RCH-20BGII	NCEH-3R
ABCH-1165Bt.	NSPL-666	Ankur-5642	RCH-2	NCEH-13Bt.
ABCH-1220Bt.	NSPL-405	Ankur-10122	RCH-530	NCEH-34 Bt.
ABCH-901-1Bt.	NSPL-999	Atal BGII	RCH-533	SBCH-310Bt.
ACH-1Bt.	Ole	Brahma	RCH-596	SBCH-292Bt.
ACH-21-1Bt.	PCH-1410Bt	GK-218BGII	SARJU BGII	TPHCNO7-015
ACH-33-1Bt.	PCH-1411Bt.	GK-221BGII	SOLAR-66BGII	TPHCNO7-005
ACH-155-1Bt.	PCH-	GK223BGII	SOLAR-60BGII	TPHCNO7-009
Akka	PCH-1412Bt.	GK-224BGII	SP-1171B2	UPLHH-189Bt.
Ankur-238Bt.	PCH-1413Bt.	GK-231 BGII	SP-504B2	UPLHH-7 Bt.
Ankur-3082Bt.	PCH-115	GK-235 BGII	SP-911B2	UPLHH-295 Bt.
Ankur-HB1024Bt.	P C H - 2 0 7 [PCH205]	GK-217	SP--904B2	UPLHH-355Bt.
Ankur-3042Bt.		Jai BG II	SP-1037	UPLHH-358Bt.
Ankur-HB1902Bt.	PCH- 409Bt.	KCH-707 BGII	Sudarshan BGII	UPLHH-360Bt.
Ankur-HB1976Bt.	PCH-930	KCH-14K59 BGII	Super-5 BGII	UPLHH-347Bt.

Bollgard I Hybrids		Bollgard II Hybrids		Other Hybrids
Brahma	PCH-2270	KCH-15K39 BG II	SWCH-2BGII	UPLHH-265Bt.
Dyna	PRCHB-405	KCH-36BGII	SWCH-4708BGII	UPLHH-271Bt.
GK-207	RCH-2	KCH-99BG II	SWCH-4703BGII	UPLHH-10Bt.
GK-209	RCH-20	KCH-135Bt.	SWCH-4715BGII	UPLHH-12Bt.
Jai Bt.	RCH-111	KDCHH-541BG II	SWCH-4720BGII	UPLHH-5Bt.
KCH-135	RCH-371	KDCHH-407BG II	SWCH-5017BGII	YRCH-4Bt.
KCH-707	RCH-368	KDCHH-441	SWCH-5011BGII	YRCH-9Bt.
MahasangramBG	RCHB-708	KDCHH-621	Tulasi-135BHII	YRCH-31Bt.
KDCH-507BG I	Rudra	KDCHH-9632	Tulasi-144 BGII	YRCH-45Bt.
KDCHB-407	Sigma	MLBC-6BG II	Tulasi-252BHII	YRCH-54Bt.
KDCH-9632	SP-1170B1	MLBCH-318	Tulasi-4 BGII	ZCH-50072Bt.
KDCH-9810	SP-1016B1	MRC-7373 BGII	Tulasi-45BHII	JKCH-1305Bt.
MECH-162	SP—911B1	MRC-7383 BGII	Tulasi-117 BGII	JKCH 229Bt.
MECH-184	SP-503	MRC-7160	Tulasi-333BHII	JK Durga
MRC-6322	SP-504(Dhanno)	MRC-7918	Tulasi-7	JKCH 99Bt.
MRC-6918	SP-700	MRC-7201	Tulasi-9	JK Ishwar
NCS-1911Bt.	SWCH-4428Bt.	MRC-7347	Tulasi-118	JKCH-634
NCS-1912Bt.	SWCH-4531Bt.	MRC-7351	VBCHB-1525BGII	JKCH-2245Bt
NCS-1913Bt.	SWCH-4314Bt.	MRC-7929	VBCHB-1526BGII	JK Chamundi Bt.
NCS-1914Bt.	Tulasi-9Bt.	NAMCOT-612	VBCH-1511BGII	JK Indira Bt.
NCS-1459Bunny)	Tulasi-4	NAMCOT-607	VBCH-1516BGII	JK Gowri Bt.
NCS-207(Mallika)	Tulasi-45Bt.	NAMCOT-604 BGII	VBCH-1519BGII	PCH-99Bt.
NCS-913	Tulasi-117	NAMCOT-605BG II	VBCH-1520BGII	PCH-77Bt.
NCS-929	Tulasi-118		VBCH-1521BGII	PRCH712Bt.
NCS-950	VBCHB-1010BG	NAMCOT-614BGII	VBCH-1501	PRCH-713Bt.
NCS-954	VBCH-1016Bt.	NAMCOT-615BGII	VBCH-1505	PRCH-714Bt.
NCS-906Bt.	VBCH-1018Bt.	NCS-854	VBCH-1506	PRCH-715Bt.
NCS-907Bt.	VBCHB-1203	NCS-207	VICH-301BGII	MH-5125Bt.
NCS-908Bt.	VICH-5	NCS-145	VICH-303BGII	MH5174Bt.
NCS-909Bt.	VICH-9	NSPL-432BG II	VICH-304BGII	BNBt. (Variety)
NCS-910Bt.	VCH-111	NSPL-333BG II	VICH-311BGII	
NCS-Bt.	118Bt.	NSPL-405	VICH-312BGII	
NCHB-940Bt.	340Bt.	NSPL-999	VICH-313BGII	
	6188Bt.	PCH-884Bt.2	VICH-314BGII	
		PCH-887Bt.2	VICH-5Bt.	
		PCH-888Bt.2	VICH-15Bt.	
		PCH-115Bt.2	110-2	

Bollgard I Hybrids		Bollgard II Hybrids		Other Hybrids
		PCH-881Bt.2	118-2	
		PCH-882Bt.2	61888-2	
		PCH-885Bt.2	322-2	
		PCH-886Bt.2	113-2	
		PCH-205Bt.2	340-2.	
		PCH-2171Bt.2		

APPENDIX 6: Company-wise and Event-wise List of *Bt*-Cotton Hybrids Approved by Government up to May 2009

Sl.No.	Name of Hybrids	Name of Company	Gene/Event	Zone
1	ABCH- 3083 Bt, ABCH3483 Bt, ABCH-1857 *Bt.*	M/s. Amar Biotech Ltd.	*cry1Ac* (MON 531)	North and Central
2	ABCH- 172 Bt, ABCH-173 Bt, ABCH- 174 Bt, ABCH- 177 Bt, ABCH-178 *Bt.*	M/s. Amar Biotech Ltd.	*cry1Ac* (MON 531)	North, Central and South
3	ABCH- 1299 *Bt.* (BG-II), ABCH- 2099 *Bt.* (BG-II), ABCH-4899 *Bt.* (BG-II), ABCH- 7399 *Bt.* (BG-II), ABCH- 143 *Bt.* BG-II, ABCH-146 *Bt.* BG-II	M/s. Amar Biotech Ltd.	*cry1Ac and cry2Ab* (MON 15985)	North and Central
4	ABCH- 1020 *Bt.* BG-II	M/s. Amar Biotech Ltd.	*cry1Ac and cry2Ab* (MON 15985)	Central
5	ABCH- 143 *Bt.* BG-II, ABCH-146 *Bt.* BG-II, ABCH- 147 *Bt.* BG-II, ABCH- 148 *Bt.* BG-II, ABCH- 1299 *Bt.* BG-II, ABCH-7399 *Bt.* BG-II	M/s. Amar Biotech Ltd.	*cry1Ac and cry2Ab* (MON 15985)	South
6	ABCH- 181 *Bt.* BG-II, ABCH-182 *Bt.* BG-II, ABCH- 191 *Bt.* BG-II, ABCH- 192 *Bt.* BG-II	M/s. Amar Biotech Ltd.	*cry1Ac and cry2Ab* (MON 15985)	North, Central and South
7	Jai Bt, Ankur 3028 Bt, Ankur 8120 *Bt.*	M/s. Ankur Seeds Pvt. Ltd	*cry1Ac* (MON 531)	North
8	Ankur 3042 *Bt.*	M/s. Ankur Seeds Pvt. Ltd	*cry1Ac* (MON 531)	Central
9	Ankur-238 Bt, Ankur 3082 Bt, Ankur HB 1024 *Bt.*	M/s. Ankur Seeds Pvt. Ltd	*cry1Ac* (MON 531)	South
10	Jai BG-II, Ankur- 3028 BG-II	M/s. Ankur Seeds Pvt. Ltd	*cry1Ac and cry2Ab* (MON 15985)	North
11	Jai BG-II, Ankur-3028 BG-II, Ankur - 3034 BGII, Ankur-216 BG-II, Ankur- 257 BG-II, Ankur- 3070 BG-II, Ankur HB 2104 BG-II	M/s. Ankur Seeds Pvt. Ltd	*cry1Ac and cry2Ab* (MON 15985)	Central
12	Jai BG-II, Ankur- 3028 BG-II, Ankur - 3034 BGII, Ankur-257 BG-II, Ankur- 356 BG-II, Ankur- 3066 BG-II, Ankur HB 2110 BG-II	M/s. Ankur Seeds Pvt. Ltd	*cry1Ac and cry2Ab* (MON 15985)	South
13	SP 7007B1	M/s. Bayer Biosciences Pvt. Ltd	*cry1Ac* (MON 531)	North
14	SP 1136 B1	M/s. Bayer Biosciences Pvt. Ltd	*cry1Ac* (MON 531)	Central

Sl.No.	Name of Hybrids	Name of Company	Gene/Event	Zone
15	SP 1170 B1, SP1016 B1, SP911B1	M/s. Bayer Biosciences Pvt. Ltd	*cry1Ac* (MON 531)	South
16	SP1169B2, SP 7010B2	M/s. Bayer Biosciences Pvt. Ltd	*cry1Ac and cry2Ab* (MON 15985)	North
17	SP 1016 B2, SP 1170 B2, SP 904B2	M/s. Bayer Biosciences Pvt. Ltd	*cry1Ac and cry2Ab* (MON 15985)	Central
18	SP-1171 B2, SP 504 B2 (Dhanno) BGII	M/s. Bayer Biosciences Pvt. Ltd	*cry1Ac and cry2Ab* (MON 15985)	South
19	311 *Bt.*	M/s. Bioseeds Research India Pvt. Ltd	*cry1Ac* (MON 531)	Central
20	118 *Bt.*	M/s. Bioseeds Research India Pvt. Ltd	*cry1Ac* (MON 531)	South
21	110-2, 111-2, 195-2	M/s. Bioseeds Research India Pvt. Ltd	*cry1Ac and cry2Ab* (MON 15985)	Central
22	110-2, 118-2, 6188-2	M/s. Bioseeds Research India Pvt. Ltd	*cry1Ac and cry2Ab* (MON 15985)	South
23	GK-218 BGII, GK 221 BGII	M/s. Ganga Kaveri Seeds Pvt. Ltd	*cry1Ac and cry2Ab* (MON 15985)	Central
24 -25	GK-218 BGII, GK-221 BGII, GK -223 BGII GK-224 BGII, GK-231 BGII, GK -235 BGII	M/s. Ganga Kaveri Seeds Pvt. Ltd M/s. Ganga Kaveri Seeds Pvt. Ltd.	*cry1Ac and cry2Ab* (MON 15985) *cry1Ac and cry2Ab* (MON 15985)	Central and South
26	GBCH-07 Bt, GBCH-09 *Bt.*	M/s. Green Gold Seeds Ltd	*GFM Cry1A* (*cry1Ab+cry 1Ac*)	Central
27	GBCH-04 Bt, GBCH-07 *Bt.*	M/s. Green Gold Seeds Ltd	*GFM Cry1A* (*cry1Ab+cry1Ac*)	South
28	JKCH-1950 Bt, JKCH-99 Bt, JKCH-1145 Bt, JKCH-1923 *Bt.*	M/s. JK Agri Genetics Ltd	*cry1Ac* Event-1	North
29	JK- Chamundi Bt, JK-Gowri Bt, JKCH-2245 Bt, JKCHB-229 Bt, JK-Ishwar (JKCH-634 Bt)	M/s. JK Agri Genetics Ltd	*cry1Ac* Event-1	Central
30	JKCH-1305 Bt, JKCHB- 229 *Bt.*	M/s. JK Agri Genetics Ltd	*cry1Ac Event-1*	South
31	KCH-36 BG-II, KCH999 BG-II, KCH-14K59 BGII, KCH-15K39 BGII	M/s. Kaveri Seed Co. Ltd	*cry1Ac and cry2Ab* (MON 15985)	North
32	KCH-14K59 BG-II, KCH-15K39 BG-II, KCH-36 BG-II, KCH999 BGII	M/s. Kaveri Seed Co. Ltd	*cry1Ac and cry2Ab* (MON 15985)	Central
33	KCH-707 BGII, KCH-14K59, BGII, KCH15K39 BGII, KCH-36 BGII, KCH-999 BGII.	M/s. Kaveri Seed Co. Ltd	*cry1Ac and cry2Ab* (MON 15985)	South

Sl.No.	Name of Hybrids	Name of Company	Gene/Event	Zone
34	KDCHB- 407 BG-I	M/s. Krishidhan Seeds Ltd.	*cry1Ac* (Mon 531)	Central
35-36	KDCHH- 507 BG-I KDCHH-541 BGII	M/s. Krishidhan Seeds Ltd. M/s. Krishidhan Seeds Ltd.	*cry1Ac* (Mon 531) *cry1Ac and cry2Ab* (MON 15985)	North, Central and South
37	KDCHB- 407 BG-II	M/s. Krishidhan Seeds Ltd.	*cry1Ac and cry2Ab* (MON 15985)	Central and South
38	MRC-7361 BG-II, MRC -7365 BG-II	M/s. Maharashtra Hybrid Seeds Co. Ltd	*cry1Ac and cry2Ab* (MON 15985)	North
39	MRC-7373 BGII, MRC -7383 BGII	M/s. Maharashtra Hybrid Seeds Co. Ltd.	*cry1Ac and cry2Ab* (MON 15985)	Central and South
40	SO7H878 BGII	M/s. Monsanto Holdings Pvt. Ltd.	*cry1Ac and cry2Ab* (MON 15985)	North
41	Brahma BGII, Krishna BGII, MLBCH6 BGII, Sudarshan BGII	M/s. Monsanto Holdings Pvt. Ltd	*cry1Ac and cry2Ab* (MON 15985)	Central
42	Sudarshan BGII, MLBCH6 BGII, Atal BGII	M/s. Monsanto Holdings Pvt. Ltd.	*cry1Ac and cry2Ab* (MON 15985)	South
43	SDS-27 BGII, SDS -6003 BGII, SDS234 BGII	M/s. Nandi Seeds Pvt. Ltd.	*cry1Ac and cry2Ab* (MON 15985)	North
44	NSPL-333 BGII, NSPL432 BGII, NSPL-666 BGII	M/s. Nandi Seeds Pvt. Ltd.	*cry1Ac and cry2Ab* (MON 15985)	Central
45	NSPL-432 BGII, NSPL333 BGII	M/s. Nandi Seeds Pvt. Ltd.	*cry1Ac and cry2Ab* (MON 15985)	South
46	NAMCOT 614 BGII, NAMCOT 615 BGII, NAMCOT- 603 BGII, NAMCOT- 605 BGII	M/s. Namdhari Seeds Pvt. Ltd	*cry1Ac and cry2Ab* (MON 15985)	Central
47	NAMCOT- 604 BG-II, NAMCOT-605 BG-II, NAMCOT-614 BG-II, NAMCOT-615 BG-II	M/s. Namdhari Seeds Pvt. Ltd	*cry1Ac and cry2Ab* (MON 15985)	South
48	NAMCOT-616 BGII, NAMCOT 617 BGII	M/s. Namdhari Seeds Pvt. Ltd.	*cry1Ac and cry2Ab* (MON 15985)	North
49	ACH 1050 Bt, ACH 1151 Bt, ACH 1171 *Bt.*	M/s. Navkar Hybrids Seeds Pvt. Ltd	*GFM Cry1A* (*cry1Ab+cry1Ac*)	Central
50	ACH-1005 Bt, ACH1085 Bt, ACH-1163 Bt, ACH-1177 *Bt.*	M/s. Navkar Hybrids Seeds Pvt. Ltd	*GFM Cry1A* (*cry1Ab+cry1Ac*)	North
51	NCS 1911 Bt, NCS1912 Bt, NCS- 1913 Bt, NCS- 1914 *Bt.*	M/s. Nuziveedu Seeds Pvt. Ltd.	*cry1Ac* (MON 531)	South
52	NCS-901 Bt, NCS-902 Bt, NCS-903 Bt, NCS904 Bt, NCS-905 *Bt.*	M/s. Nuziveedu Seeds Pvt. Ltd	*cry1Ac* (MON 531)	North
53	NCS-906 Bt, NCS-907 Bt, NCS-908 Bt, NCS909 Bt, NCS-910 *Bt.*	M/s. Nuziveedu Seeds Pvt. Ltd	*cry1Ac* (MON 531)	Central

Sl.No.	Name of Hybrids	Name of Company	Gene/Event	Zone
54	NCS- 852 Bt2, NCS- 859 Bt2, NCS- 860 Bt2, NCS- 861 Bt2, NCS- 862 Bt2, NCS- 863 Bt2, NCS- 864 Bt2, NCS- 865 Bt2, NCS- 866 Bt2, NCS- 867 Bt2	M/s. Nuziveedu Seeds Pvt. Ltd.	*cry1Ac and cry2Ab* (MON 15985)	South
55	NCS- 859 Bt2, NCS-860 Bt2, NCS-861 Bt2, NCS862 Bt2, NCS- 853 Bt2	M/s. Nuziveedu Seeds Pvt. Ltd	*cry1Ac and cry2Ab* (MON 15985)	Central
56	NCS- 855 Bt2, NCS-856 Bt2, NCS-857 Bt2, NCS858 *Bt.*	M/s. Nuziveedu Seeds Pvt. Ltd	*cry1Ac and cry2Ab* (MON 15985)	North
57	PCH-99 Bt, PCH-77 *Bt.*	M/s. Palamoor Seeds Pvt. Ltd.	*cry1Ac* (JK Event-1)	Central and South
58	PCH- 1410 Bt, PCH1411 Bt, PCH- 1412 Bt, PCH- 1413 *Bt.*	M/s. Prabhat Agri Biotech Ltd.	*cry1Ac* (MON 531)	South
59	PCH 401 Bt, PCH 402 Bt, PCH 403 *Bt.*	M/s. Prabhat Agri Biotech Ltd	*cry1Ac* (MON 531)	North
60	PCH-404 Bt, PCH-405Bt, PCH-407Bt, PCH-408Bt, PCH-409Bt	M/s. Prabhat Agri Biotech Ltd	*cry1Ac* (MON 531)	Central
61	PCH- 884 Bt2, PCH- 887 Bt2, PCH- 888 Bt2, PCH- 115 Bt2, PCH- 881 Bt2, PCH- 882 Bt2, PCH- 885 Bt2, PCH- 886 Bt2, PCH- 205 Bt2, PCH- 2171 Bt2	M/s. Prabhat Agri Biotech Ltd.	*cry1Ac and cry2Ab* (MON 15985)	South
62	PCH-876 Bt2, PCH-877 Bt2, PCH-878 Bt2, PCH879 Bt2	M/s. Prabhat Agri Biotech Ltd	*cry1Ac and cry2Ab* (MON 15985)	North
63	PCH-115 Bt2, PCH-881 Bt2, PCH-882 Bt2	M/s. Prabhat Agri Biotech Ltd	*cry1Ac and cry2Ab* (MON 15985)	Central
64	PRCHB-405 BGI	M/s. Prava rdhan Seeds Pvt. Ltd	*cry1Ac* (MON 531)	Central
65	PRCH-302, PRCH-333	M/s. Pravardhan Seeds Pvt. Ltd	*cry1Ac and cry2Ab* (MON 15985)	North
66	PRCH-331 *Bt.* II, PRCH333 *Bt.* II	M/s. Pravardhan Seeds Pvt. Ltd	*cry1Ac and cry2Ab* (MON 15985)	Central and South
67	PRCH- 712 Bt, PRCH-713 Bt, PRCH- 714 Bt, PRCH- 715 *Bt.*	M/s. Pravardhan Seeds Pvt. Ltd.	*cry1Ac* (JK Event-1)	Central and South
68	Shakti- 9Bt	M/s. Rasi Seeds Pvt. Ltd	*cry1Ac* (MON 531)	North
69	RCH-602 BGII, RCH605 BGII, RCH-314 BGII	M/s. Rasi Seeds Pvt. Ltd	*cry1Ac and cry2Ab* (MON 15985)	North
70	RCH-608 BGII and RCH377 BG-II, RCH- 530 BGII	M/s. Rasi Seeds Pvt. Ltd	*cry1Ac and cry2Ab* (MON 15985)	Central
71	RCH-20 BG-II	M/s. Rasi Seeds Pvt. Ltd	*cry1Ac and cry2Ab* (MON 15985)	South
72	TPHCN07-015 Bt, TPHCN07-005 Bt, TPHCN07-009 *Bt.*	M/s. RJ Biotech Pvt. Ltd.	*GFM cry1A (cry1Ab+cry1Ac)*	Central and South

Sl.No.	Name of Hybrids	Name of Company	Gene/Event	Zone
73	SBCH- 286 *Bt.* (Raka Bt)	M/s. Safal Seeds and Biotech Ltd	*GFM Cry1A (cry1Ab+cry1Ac)*	Central
74	SBCH- 310 *Bt.*	M/s. Safal Seeds and Biotech Ltd.	*GFM cry1A (cry1Ab+cry1Ac)*	South
75	SWCH- 4428 Bt, SWCH- 4531 Bt, SWCH- 4314 *Bt.*	M/s. Seed Works International Pvt Ltd	*cry1Ac* (MON 531)	Central and South
76	SWCH-4707 BG-II, SWCH-4711 BG-II, SWCH-2 BG-II, SWCH4704 BG-II, SWCH-4713 BG-II	M/s. Seed Works International Pvt Ltd	*cry1Ac and cry2Ab* (MON 15985)	North
77	SWCH-2 BG-II, SWCH4708 BG-II, SWCH4715 BG-II, SWCH-1 BG-II, SWCH- 5017, SWCH-5011	M/s. Seed Works International Pvt Ltd	*cry1Ac and cry2Ab* (MON 15985)	Central
78	SWCH-2 BG-II, SWCH4708 BG-II, SWCH4703 BG-II, SWCH4715 BG-II, SWCH-4720 BG-II, SWCH5017 BGII, SWCH-5011 BG II	M/s. Seed Works International Pvt Ltd	*cry1Ac and cry2Ab* (MON 15985)	South
79	SOLAR- 56 BG-II, SOLAR- 64 BG-II, and SOLAR-65 BG-II, SOLAR-72 BG-II, SOLAR-75 BGII, SOLAR-76 BG-II, SOLAR-77 BG-II	M/s. Solar Agrotech Pvt. Ltd.	*cry1Ac and cry2Ab* (MON 15985)	North
80	SOLAR- 66 BG-II, SARJU BG-II, SOLAR60 BG-II	M/s. Solar Agrotech Pvt. Ltd	*cry1Ac and cry2Ab* (MON 15985)	Central and South
81	Super- 5 BG-II	M/s. Super Seeds Pvt. Ltd.	*cry1Ac and cry2Ab* (MON 15985)	South and Central
82	Tulasi-162 BG-II, Tulasi- 225 BG-II	M/s. Tulasi M/s. Seeds Pvt. Ltd	*cry1Ac and cry2Ab* (MON 15985)	North
83	Tulasi- 135 BG-II, Tulasi-144 BG-II, Tulasi- 162 BG-II, Tulasi-117 BG-II	M/s. Tulasi Seeds Pvt. Ltd	*cry1Ac and cry2Ab* (MON 15985)	Central
84	Tulasi- 135 BG-II, Tulasi-144 BG-II, Tulasi- 252 BG-II, Tulasi- 4 BG-II, Tulasi45 BG-II, Tulasi- 117 BG-II, Tulasi- 333 BG-II	M/s. Tulasi Seeds Pvt. Ltd.	*cry1Ac and cry2Ab* (MON 15985)	South
85	UPLHH-12 Bt, UPLHH271 *Bt.*	M/s. Uniphos Enterprises Ltd.	*GFM cry1A (cry1Ab+cry1Ac)*	North
86	UPLHH-17 Bt, UPLHH12 Bt, UPLHH- 271 Bt, UPLHH-1Bt, UPLHH-10 *Bt.*	M/s. Uniphos Enterprises Ltd.	*GFM cry1A (cry1Ab+cry1Ac)*	Central
87	UPLHH- 360 Bt, UPLHH- 347 *Bt.* and UPLHH- 265 Bt, UPLHH- 271 *Bt.* and UPLHH- 10 *Bt.*	M/s. Uniphos Enterprises Ltd.	*GFM cry1A (cry1Ab+cry1Ac)*	South
88	UPLHH-342 Bt, UPLHH-350 *Bt.*	M/s. Uniphos Seeds and Biogenetics	*GFM cry1A (cry1Ab+cry1Ac)*	North

Sl.No.	Name of Hybrids	Name of Company	Gene/Event	Zone
89	UPLHH-189 Bt, UPLHH-352 Bt, UPLHH- 13 Bt.	Uniphos Seeds and Bio-Genetics	GFM cry1A (cry1Ab+cry1Ac)	Central
90	UPLHH- 189 Bt. and UPLHH-7 Bt, UPLHH295 Bt, UPLHH-355 Bt. and UPLHH- 358 Bt.	Uniphos Seeds and Bio-Genetics	GFM cry1A (cry1Ab+cry1Ac)	South
91	VBCH 1515 BGII, VBCH1516 BGII, VBCH-1517 BGII, VBCH-1518 BGII	M/s. Vibha Agrotech Ltd	cry1Ac and cry2Ab (MON 15985)	North
92	VBCH-1511, VBCH1516, VBCH-1519, VBCH-1520, VBCH1521,VBCHB- 1525, VBCHB- 1526	M/s. Vibha Agrotech Ltd	cry1Ac and cry2Ab (MON 15985)	Central
93	VBCH-1511 BG-II, VBCH-1516 BG-II, VBCH-1519 BG-II, VBCH-1520 BG-II, VBCH-1521 BG-II, VBCHB- 1525 BG-II, VBCHB-1526 BG-II	M/s. Vibha Agrotech Ltd.	cry1Ac and cry2Ab (MON 15985)	South
94	VICH- 301 BG-II, VICH-303 BG-II, VICH304 BG-II.	M/s. Vikram Seeds Ltd	cry1Ac and cry2Ab (MON 15985)	Central
95	VICH-307 BG-II, VICH308 BG-II, VICH-309 BG-II, VICH-310 BG-II	M/s. Vikram Seeds Ltd	cry1Ac and cry2Ab (MON 15985)	North
96	VICH- 301 BG-II, VICH-303 BG-II, VICH304 BG-II, VICH-311 BG-II, VICH- 312 BG-II, VICH- 313 BG-II, VICH- 314 BG-II	M/s. Vikram Seeds Ltd.	cry1Ac and cry2Ab (MON 15985)	South
97	YRCH-4 Bt, YRCH-9 Bt, YRCH-13 Bt, YRCH-31 Bt, YRCH-45 Bt, YRCH-54 Bt.	M/s. Yashoda Hybrid Seeds Ltd.	GFM Cry1A (cry1Ab+cry1Ac)	Central and South
98	ZCH-193 Bt.	M/s. Zuari Seeds Ltd	GFM Cry1A (cry1Ab+cry1Ac)	North

APPENDIX 7: Name and Address of some Seed Companies of India

1. **Advanta India Ltd.**, 203-205, 2nd Floor Bhuvana Towers, S.D. Road, Secunderabad - 500 003.

2. **AgSun Seeds (India) Pvt. Ltd**. 603, Muralidhar Chambers, 352 JSS Road, Thakurdwar, Mumbai-400 002.

3. **Ajeet Seeds** Ltd., Tapadia Terraces, 2nd Floor, Adalat Road, Aurangabad-431 001.

4. **Amareswara Agri-Tech Ltd.**, 4th Floor, Camus Capri Apartment, Rajbhavan Road, Somajiguda, Hyderabad-500 082.

5. **Ankur Seeds Pvt. Ltd.**,27, New Cotton Market Lay-out, Nagpur- 440 018.

6. **Avani Seeds Ltd.**, 5-8, Ashoka Chambers, II Floor, Near Lions Hall, Mithakhali Six Roads, Ellisbridge, Ahmedabad-380 006.

7. **Basant Agro Tech India Ltd.**, Near S.T. Workshop, Kaulkhed, Akola-444 004.

8. **Bayer Bioscience Pvt. Ltd**. Dhumaspur Road, Badshahpur, Gurgaon – 122 001.

9. **Bejo Sheetal Seeds Pvt. Ltd.**, PO Box 77, Bejo Sheetal Corner Mantha Road, Jalna -431 203 (MS).

10. **Bharathi Seeds Ltd.**, 5L/28, Opp.Petrol Pump Noonepalli, Nandyal - 518 503 (AP). Srinivasanagar, Nandyal.

11. **Biostadt MH seeds Ltd.**, B-24 and W-4, MIDC Area, Chikalthana, Aurangabad - 431 210 (MS).

12. **Century Seeds Pvt. Ltd.**,BA-22-24, Phase II, Mangolpuri Industrial Area, New Delhi -110 034.

13. **Devgen Seeds and Crop Technology Pvt. Ltd.**, Flat No.101, Sapthagiri Residency, 1-10-98/A, Chikoti Gardens, Begumpet, Hyderabad - 500 016.

14. **Doctor Seeds Pvt. Ltd.**, Room No.5, 5th Floor, Carnival Shopping Complex, The Mall, Ludhiana-141 001 (Punjab).

15. **ECL Agrotech Ltd.**, No. 101, Varsha Apartment, 221/224, Sir C.V. Raman Road, RMV Extension Bangalore - 560 080.

16. **Green Gold Seeds Ltd.**, 1st Floor, Shyam Chambers, Bansilal Nagar, Railway Station Road, Aurangabad-431 005.

17. **Hash Biotech Labs Pvt. Ltd.**, SCO 311-312, Sector 40-D Chandigarh - 160 039.

18. **Hi-Gene Seeds (India) Pvt. Ltd.**, Plot No.68, Jaya Nagar, New Bowenpally, Secunderabad - 500 011.

19. **Hi-tech Agro Enterprises,** Plot No.M-5 and 6, Industrial Estate, Hatalgeri Naka, Gadag - 582 101 (Karnataka).

20. **Hindustan Lever Ltd.**, 165/166, Backbay Reclamation, Mumbai -400 020.

21. **Indbro Research and Breeding Farms Pvt. Ltd.,** 302, Sri Sai Krishna Resideny, Opp. NTR Stadium, Arvind Nagar, Domalguda, Hyderabad-500 029.

22. **Indo-American Hybrid Seeds (India) Pvt. Ltd.**, 7th Kilometer Banashankari-Kengeri Link Road Channasandra Village, Uttarahalli Hobli, Subramanyapura Post, Bangalore -560 061.

23. **JK Agri Genetics Limited**, 1-10-177, 4th Floor, Varun Towers, Begumpet, Hyderabad-500 016.

24. **Kaveri Seed Company (P) Limited**, 513 B, V Floor, Minerva Complex, S.D. Road, Secunderabad – 500 003.

25. **Krishidhan Seeds Ltd.**, P.B.No.92, 34-C, New Mondha, Jalna -431 203 (Maharashtra).

26. **Maharashtra Hybrid Seeds Co. Ltd.**, Resham Bhavan, 4th Floor, 78, Veer Nariman Road, Mumbai - 400 020.

27. **Mahodaya Hybrid Seeds Pvt. Ltd.**, 1-34-214, Head Post Office Road, Jalna - 431 203, Maharashtra.

28. **Markfield Hybrid Seeds (Pvt.)Ltd.**, B-6, Pawar Chambers, Bus Stand Road, Jalna - 431 203 (Maharashtra).

29. **Monsanto Genetics India Pvt. Ltd.**, 5th Floor, Ahura Centre, 96 Mahakali Caves Road, Andheri (East), Mumbai - 400 093.

30. **MR Seeds (P) Ltd.**, 82-A, Old Dhan Mandi Shri Ganga Nagar – 335 001 (Raj.).

31. **Namdhari Seeds Pvt. Ltd.** 119, 9th Main Road, Ideal Homes, Rajarajeswari Nagar, Bangalore -560 098.

32. **Nandi Seeds Private Limited**, 1-7-44/A/1, Market Road, Mahabubnagar – 509 001 (Maharashtra).

33. **Narmada Agritech Pvt. Ltd.**, Narmada House, Vijayashanti Enclave, Kompally, Secunderabad – 500 014.

34. **Nath Bio-Genes (India) Ltd.**, Nath House, Nath Road, Aurangabad-431 005 (M.S.).

35. **Navkar Hybrid Seeds Pvt. Ltd.**, Bakar Ali's Wadi, Mirzapur Char Rasta, Ahmadabad - 380 001.

36. **New Nandi Seed Corporation**, 1, Patal Society, Near Sears Tower, Opp.-Inklab Society, Gulbai Tekra, Ellis Bridge, Ahmedabad-380 006.

37. **Nimbkar Seeds Pvt. Ltd.**,Jinti Bridge, Phaltan-415 523, Distt. Satara (M.S.).

38. **Nirmal Seeds Pvt. Ltd.**, Bhadgaon Road, Pachora -424 201 Dist. Jalgaon, (Maharashtra).

39. **Noble Seeds Pvt. Ltd.**, 33, Noble House, Khera Kalan, Delhi - 110 082.

40. **Nu Genes Pvt. Ltd.**, (formerly Nitya Seed Sciences Pvt. Ltd.), Plot No.18, 1st Floor, NCL Enclave, Jeedimetla, NH-7, Medchal Road, Secunderabad-500 055 (A.P.).

41. **Nunhems Seeds Pvt. Ltd.**, Dhumaspur Road, Badshahpur, Gurgaon-122 001.

42. **Nuziveedu Seeds Limited**, 7th Floor, C- Block, 104,Surya Towers, S.P.Road, Secunderabad - 500 003.

43. **Palamoor Seeds Pvt. Ltd.**, K-1, Municipal Complex, New Town, Mahabubnagar - 509 001 (A.P.).

44. **PHS Agritech (P) Ltd.**, Plot 3, Laxmi Nagar Colony, Behind Hanuman Temple, Picket, Secunderabad – 500 009.

45. **Plantgene Seeds Limited**, No.4-170/1, Plot No.72, Venkataramana Colony, Loyola Academy Road, Old Alwal, Secunderabad - 500 010.

46. **Prabhat Agri Biotech Ltd.**, 6-3-540/10, Opp.SBH, Punjagutta, Hyderabad – 500 082.

47. **Pravardhan Seeds Pvt. Ltd.**, 3-4-754, Baghlingampally, Barkatpura Main Road,Hyderabad - 500 027.

48. **Safal Seeds and Biotech Ltd.**,A-2, Old M.I.D.C., Jalna-431 203 (Maharashtra).

49. **Senthil Seeds,** 73, Karur Road Kolathupalayam, Dharapuram Tk. Erode Distt. (TN). Dharampuram (Distt. Erode).

50. **Hybrid Seeds Pvt. Ltd.**, Tilak Bazar, Near Bus Stand, Hisar – 1250 01.

51. **Shri Ram Seeds,** 38, Jawahar Market, Sriganganagar - 335 001 (Raj.)

52. **Rama Agri Genetics (India) Pvt. Ltd.**, H.No.11-68, Ground Floor-2, Siddeswara Apartments, Krishna Nagar, Kurnool-518 002 (A.P.).

53. **Sri Sathya Agri Biotech Pvt. Ltd.**, Door No.8-2-103/1, Shaktinagar, Sagar Road, L.B. Nagar, Hyderabad - 500 074.

54. **Sumitra Seeds Pvt. Ltd.**, D.No.8-148, Plot No.31, Industrial Estate, Kurnool – 518003.

55. **Sun Agrigenetics P. Ltd.**, Reign Plaza, 2nd Floor, Near GEB Substation, Gotri Road, Baroda – 390 021.

56. **Super Agri Seeds Pvt. Ltd.**,No.3-7-230, First Floor, Vikrampuri Colony, Secunderabad - 500 009.

57. **Super Seeds Pvt. Ltd.**, 32, New Anaj Mandi, Hisar – 125001.

58. **Surya Seeds Ltd.**, (Formerly Surya Seeds Pvt. Ltd.), 6-20-10, Sivaraj Arcade, 7/1, Arundel pet, Guntur - 522 002.

59. **Swagath Seeds Pvt. Ltd.**, 3-6-290/21, II Floor, Sadhana Building, Hyderguda, Hyderabad – 500 029.

60. **Syngenta India Ltd.**, Royal Insurance Building, 14, J.Tata Road, Churchgate, Mumbai-400020.

61. **Tulasi Seeds Pvt. Ltd.**, 'Tulasi House', 6-4-6, 4/5 Arundelpet, Guntur - 522 002.

62. **Unicorn Seeds Ltd.**, 1-7-139/3. S.D. Road, Secunderabad-500 003.

63. **United Genetics India Pvt. Ltd.**, No.527, "F" Block, 60 Feet Road, Sahakarnagar, Bangalore - 560 092.

64. **Vibha Agrotech Ltd.**, 501, Subhan Sirisampada Complex, Rajbhavan Road, Somajiguda, Hyderabad-500 082.

65. **Vikki's Agrotech Ltd.**, H.No. 8-3-960/6/2, Srinagar Colony, Hyderabad 500 073.

66. **Vikram Seeds Ltd.**, 209, Ashwamegh Avenue; Nr. Mithakhali, Underbridge, Mayur Colony; avrangpura, Ahmedabad-380 009.

67. **Vishal Seeds (P) Ltd.**, 8-2-108/3, Hasthinapuram North Nagarjuna Sagar Road, Hyderabad-500 070.

68. **VNR Seeds Pvt. Ltd.**, Shop No.1, First Floor, Durga College Complex, K.K. Road, Moudhapara, Raipur-492 001.

69. **Zuari Seeds Limited**, No. 805, 13th A Main, 80 Feet Road, Yelahanka New Town, Bangalore - 560 064.

Glossary

A Line: The male sterile line.

B. Line: Isogenic line of A line which is male fertile.

Bale: The unit of weighing cotton. In India one bale is equal to 170 kg.

Boll: The fruit of cotton bearing seed cotton.

Burs: Woody structures of boll rind which remain on the plant after picking.

Breeders' Exemption: Rights provided to use protected material as the basis to develop a new variety or for other research work; also called research exemptions or breeders' privilege.

Candidate Variety: A variety to be registered under Plant Variety Protection Act is referred to as candidate variety.

Cotton: Those species of the genus Gossypium which bear spinnable seed coat fibres.

Colored Cotton: Cotton with other than white colour of the lint.

Container: A box, bottle, casket, tin, barrel, case, receptacle, sack, bag, wrapper or other thing in which any article or thing is placed or packed.

Dealer: A person who carries on the business of buying and selling, exporting, or importing seed, and includes an agent.

Delinting: The process of removing fuzz from cotton seed.

Distinctiveness: The variety must be distinguishable in atleast one character from previously available varieties.

Egyptian cotton: *Gossypium barbadense;* also called Peruvian cotton or Tanguish cotton or Sea Island cotton.

Essentially derived variety: A variety derived from the initial variety and having most of the characteristics of initial variety but clearly distinguishable from initial variety.

Example Variety: A variety that is used for comparison for a particular character is called example variety.

Extant Variety: All released, notified and unprotected varieties. In other words, various varieties which are available in India such as farmers' variety, variety of common knowledge or any other variety of public domain.

Farmer: Any person who cultivates crops either by cultivating the land himself or through any other person but does not include any individual, company, trader or dealer who engages in the procurement and sale of seeds on a commercial scale.

Farmers' Exemption: Legal rights provided to farmers to save, use, exchange, share or sell his farm produce of a protected variety; also called farmers' rights or farmers' privilege.

Farmers' Rights: Rights arising from the past, present and future contributions of farmers in conserving, improving and making available plant or animal genetic resources.

Farmers' Variety: A variety that has been developed by a farmer and used for commercial cultivation for several years is called farmers variety.

Frego Bract: The narrow or twisted type of bract.

Fuzz: Short non-spinnable seed coat fibres,

Gin: A device which is used for separation of lint from seed.

Ginning: The process of separation of lint from the seed.

Ginning per cent: The quantity of lint obtained from 100 grams of seed cotton.

Hank: The yarn of 840 yard length.

Hull: The outer most covering of seed bearing lint and fuzz.

Hull Index: The weight of hull in gram obtained from 100 seeds.

Kidney Cotton: The race Brasilience of *Gossypium barbadense*.

Kind: One or more related species or sub-species of crop plants each individually or collectively known by one common name such as cabbage, maize, paddy and wheat.

Lint: Long spinnable seed coat fibres.

Lea: The yarn of 120 yard length.

Lint Index: Weight of lint in gram obtained from 100 seeds.

Misbranded: A seed with false label or label of another variety.

Monopodia: Vegetative branches which do not bear boll directly.

Motes: Sterile, abortive and abnormal ovules.

Naked seed: Seeds with smooth surface after ginning.

Nectaries: Nector containing glands found on leaf, bracts and base of calyx.

Novelty: The newness of a variety means not previously sold.

Okra Leaf: Cotton genotypes with narrow lobed leaves.

Old World Cottons: Diploid cultivated species of cotton [*G. arboreum* and *G. herbaceum*].

Plant Breeders' Rights: Plant breeders' rights [PBR], also known as plant variety rights (PVR), are intellectual property rights granted to the breeder of a new variety of plant.

Producer: A person, group of persons, firm or organization who grow or organize the production of seeds.

R Line: Line which restores fertility when crossed with cytoplasmic genic male sterile line.

Reference Variety: All released and notified extant varieties of common knowledge which are in seed production chain.

Seed: Any type of living embryo or propagule capable of regeneration and giving rise to a plant of agriculture which is true to such type.

Seed Cotton: Raw cotton obtained after picking.

Seed Export: Taking seed out of India by land, sea or air.

Seed Import: Bringing seed into India by land, sea or air.

Seed Index: Weight of 100 cotton seeds in gram.

Seed Processing: The process by which seeds and planting materials are dried, threshed, shelled, ginned or delinted (in cotton), cleaned, graded or treated;

Spurious Seed: Any seed which is not genuine or true to type.

Square: Unopened flower bud.

Spinning: The process of making yarn from the lint.

Stalk: The woody portion of cotton plant which remains in the field after final picking.

Stability: Each generation of the variety should be similar.

Sympodia: Boll bearing branches.

Tetraploid Cottons: *Gossypium hirsutum* and *G. barbadense.*

Transgenic Variety: Seed or planting material synthesized or developed by modifying or altering the genetic composition by means of genetic engineering.

Trash: Foreign material in the seed cotton.

Tree Cotton: *Gossypium arboreum* cotton.

Uniformity: All plants in a variety should look alike.

Upland Cotton: *Gossypium hirsutum* cotton; also called American cotton.

Variety: A genotype released for commercial cultivation either by State Variety Release Committee or by Central Variety Release Committee.

Zero Branch: Cotton genotypes bearing bolls directly on the main stem.

References

1. Allison, D.C. and Fisher, W.D. 1964. A dominant gene for male sterility in upland cotton. *Crop Sci.,* 4 : 548-549.

2. Babu, V.R., Punit Mohan, Singh, P. and Singh, V.V. 2003. Salt tolerance in the germplasm of diploid cottons. *Indian J. Agric. Sci.* 73 : (5): 301-302.

3. Basu, A. K., Mahendra Singh, Kumudini Nautyal. 1999. Sankar Kapas [Hybrid Cotton], Indian Council of Agricultural Research, New Delhi.

4. Basu, A.K., Singh, Phundan and Narayanan, S. S. 1993. Progress of Breeding barbadense cotton in India. *ISCI Journal*, 18(2): 95-102.

5. Bowman, D.T. and Weaver, J.B. Jr. 1979. Analysis of a dominant male sterile character in upland cotton II. Genetic Studies. *Crop Sci.,* 19 : 628-630.

6. CICR, 1989. A Catalogue of Cotton Genetic Resources in India. Published by CICR, Nagpur.

7. Gotmare V. and Singh P. 2004. Use of wild Species for Cotton Improvement in India ICAC Recorder, 22 (3): 12-14, Sept. 2004.

8. Hutchinson, J. B., Silow, R. A. and Stephens, S. G. 1947. The Evolution of Gossypium and the differentiation of the cultivated cotton. Oxford Uni. Press, London.

9. Justus, N.J. and Lienweber, C.L. 1960. A heritable partial male sterile character in cotton. *J. Hered.* 51: 191-192.

10. Justus, N.J., Meyer, J.R. and Roux, J.B. 1963. A partially male sterile character in upland cotton. *Crop Sci.,* 3: 428-429.

11. Kohel, R. J. and Lewis, C. F. (eds) 1984. Cotton Monograph 24. ASA/CSSA/SSSA, Madison, USA.

12. Mayee, C.D., Singh, P., Punit Mohan and D.K. Agarwal 2004. Evaluation of *Bt.* transgenic intra-*hirsutum* hybrids for yield and fibre properties. *Indian J. Agric. Sci.* 74 (1): 46-47.

13. Mayee, C.D., **Singh, Phundan**, Dongre, A.B., Rao, M.R.K. and Sheo Raj 2002. Transgenic *Bt.* cotton. CICR, Technical Bulletin No. 22.

14. Meshram, L.D., Ghongade, R.A. and Marawar, M.W. 1994. Development of male sterility systems from various sources in cotton. *PKV Res. J.* 18 (1): 83-86.

15. Meyer,V.G. 1975. Male sterility from *Gossypium harknessii. J. Hered.* 66 : 23-27.

16. Muller, G. 1968. Cotton: Series of Monograph on Tropical and Subtropical Crops. Ruhr-Stickstoff Aktiengesellschaft Bochum, West Germany.

17. Munro, J.M. (ed) 1987. Cotton. Longman Group Group Limited, UK.

18. Niles, G. A. 1980. Plant Breeding and Improvement of Cotton Plant. Outlook on Agriculture, 10:152-158.

19. Percy, R.G. and Turcotte, E.L. 1991. Inheritance of male sterile mutant ms$_{13}$ in American Pima Cotton. *Crop Sci.* 31 (6) : 1520-1521.

20. Richmond, T.R. and Kohel, R.J. 1961. Analysis of a completely male sterile character in American upland cotton. *Crop Sci.* 1 : 397-401.

21. Rhyne, C.L. 1971. Indehisent anther in cotton. *Cotton Gr. Rev.* 48: 194-199.

22. Santhanam, V. 1976. Cotton. ICAR Low Priced Publication.

23. Sethi, B.L., Sikka, S.M., Dastur, R.H., Gadkari, P.D., Balasubramanian, R., Maheshwari, P., Rangaswamy, N.S. and Joshi, A. B. 1960. Cotton in India: A Monograph, Published by Indian Central Cotton Committee, Mumbai.

24. Silow, R.A. 1944. Genetics of species development in Old World Cotton. J. Genet. 66:62-67.

25. Singh, D.P. and Kumar, R. 1993. Male genetic sterility in Asiatic cotton. *Indian J. Genet.* 53 (1): 99-100.

26. **Singh, Phundan**. 2012. Cotton Breeding 3rd edn. Kalyani Publishers, New Delhi.

27. **Singh, Phundan** and Singh, Sanjeev. 2010 Breeding Hybrid Cotton. 3rd edn. Kalyani Publishers, New Delhi.

28. **Singh, Phundan**. 2012. Breeding Transgenic *Bt.* Cotton. Kalyani Publishers, New Delhi.

29. **Singh, Phundan**. 2012. Cotton Improvement in India. New Vishal Publications, New Delhi.

30. **Singh, Phundan** and Narayanan, S. S. 1991.Genetical Improvement of *arboreum* cotton in India. ISCI Journal, 16(2):81-96.

31. **Singh, Phundan** and Narayanan, S. S. 1999. Cotton Improvement Procedures. In Hand Book of Cotton in India, Published by ISCI, Mumbai.

32. **Singh, Phundan** and Narayanan, S. S. 1993 Breeding of herbaceum cotton in India. J. Cott. Res. Dev., 7(1):1-8.

33. **Singh, Phundan** and Kairon, M.S. 2000. Cotton Varieties and Hybrids. CICR Technical Bulletin No. 13.

34. **Singh, Phundan**, Kairon, M.S. and Singh S.B. 2000. Breeding Hybrid Cotton. CICR Technical Bulletin No. 14.

35. **Singh, Phundan.**, Basu, A.K., Sundaram, V. and Punit Mohan 2003. History of Cotton Breeding in India. ISCI Journal, 28 (1): 1-7.

36. **Singh, Phundan** and Mayee, C.D. 2001. Role of hybrids in cotton improvement and Indian economy. The Botanica, 51 : 1-8. 2001. Delhi University Journal.

37. **Singh, Phundan** and Mayee, C.D. 2001. Present status and future prospects of hybrid cotton in India. FAO Hybrid Cotton Newsletter, Vol. 10, 2001.

38. Singh, **Phundan** and Punit Mohan 2005. Progress and prospects of R and D in diploid (*desi*) cottons in India. ISCI Journal, 30(2): 75-84.

39. Singh, S.B., **Singh, Phundan** and Mayee, C.D. 2002. Male sterility in cotton. CICR, Technical Bulletin No. 24.

40. Turcotte, E.L. and Feaster, C.L. 1979. Linkage tests in American Pima Cotton. *Crop Sci.* 19 : 119-120.

41. Turcotte, E.L. and Feaster, C.L. 1985. Inheritance of male sterile mutant Ms_{12} in American Pima Cotton (*Gossypium barbadense*). *Crop Sci.* 25 (4): 688-690.

42. Tuteja, O. P., Verma, S. K., Monga, D. and **Singh, P.** 2005. A new genetic male sterile line of *desi* cotton (*Gossypium arboreum* L.). Indian J. Genet. 65 (2) : 145-146.

43. Valicek, P. 1978. Wild and Cultivated Cottons. Coton et Fibres Tropicales, 33:363-387.

44. Weaver, J.B. Jr. 1968. Analysis of a double recessive completely male sterile cotton. *Crop Sci.* 8: 597-600.

45. Weaver, J.B. Jr. and Ashley, T. 1971. Analysis of a dominant gene for male sterility in upland cotton *Gossypium hirsutum. Crop Sci.* 11 : 596-598.

46. Zhang, T.Z., Yijun, F. and Zjiyaju, P. 1994. Genetic evaluation of genetic male sterile lines found in People Republic of China in *Gossypium hirsutum* L. Proceedings: World Cotton Conference-I. February 13-17, Brisbane, Australia (1994).

Books by the Same Author

A. PLANT BREEDING

1. Essentials of Plant Breeding
2. Theory of Plant Breeding
3. Plant Breeding: Molecular and New Approaches
4. Molecular Plant Breeding
5. Principles of Seed Technology
6. Seed Technology: At a Glance
7. Objective Seed Technology
8. Practical in Crop Breeding
9. Practical and Numerical in Plant Breeding
10. Numerical Problems in Plant Breeding and Genetics
11. Objective Science of Plant Breeding
12. Plant Breeding At a Glance
13. Plant Breeding [For Under Graduate Students]
14. Principles of Plant Breeding
15. Cotton Breeding
16. Heterosis Breeding in Cotton
17. Breeding Hybrid Cotton
18. Breeding Transgenic *Bt.* Cotton
19. Cotton Improvement in India
20. Glimpses of Cotton Breeding
21. Sankar Kapas (Hindi)
22. Glossary of Plant Breeding and Genetics

E. INTELLECTUAL PROPERTY RIGHTS

F. COMPETITIVE EXAMINATION

Index

www.ingramcontent.com/pod-product-compliance
Lightning Source LLC
Chambersburg PA
CBHW050228270326
41914CB00003BA/624